S0-BZY-747

Linear Algebra Gems

Assets for
Undergraduate Mathematics

©2002 by the Mathematical Association of America (Inc.)

ISBN: 0-88385-170-9

Library of Congress Catalog Card Number 2001097387

Printed in the United States of America

Current Printing

10 9 8 7 6 5 4 3 2 1

Linear Algebra Gems

Assets for Undergraduate Mathematics

Edited by

David Carlson
Charles R. Johnson
David C. Lay
A. Duane Porter

Published and Distributed by
The Mathematical Association of America

The MAA Notes Series, started in 1982, addresses a broad range of topics and themes of interest to all who are involved with undergraduate mathematics. The volumes in this series are readable, informative, and useful, and help the mathematical community keep up with developments of importance to mathematics.

Editorial Board
Sr. Barbara E. Reynolds, Editor
Nancy Baxter Hastings, Associate Editor
Jack Bookman   Annalisa Crannell   Daniel John Curtin
Paul E. Fishback   Nancy L. Hagelgans   Richard D. Järvinen
Ellen J. Maycock   Mary R. Parker

MAA Notes

11. Keys to Improved Instruction by Teaching Assistants and Part-Time Instructors, *Committee on Teaching Assistants and Part-Time Instructors, Bettye Anne Case*, Editor.

13. Reshaping College Mathematics, *Committee on the Undergraduate Program in Mathematics, Lynn A. Steen*, Editor.

14. Mathematical Writing, by *Donald E. Knuth, Tracy Larrabee, and Paul M. Roberts*.

16. Using Writing to Teach Mathematics, *Andrew Sterrett*, Editor.

17. Priming the Calculus Pump: Innovations and Resources, *Committee on Calculus Reform and the First Two Years*, a subcomittee of the Committee on the Undergraduate Program in Mathematics, *Thomas W. Tucker*, Editor.

18. Models for Undergraduate Research in Mathematics, *Lester Senechal*, Editor.

19. Visualization in Teaching and Learning Mathematics, *Committee on Computers in Mathematics Education, Steve Cunningham and Walter S. Zimmermann*, Editors.

20. The Laboratory Approach to Teaching Calculus, *L. Carl Leinbach et al.*, Editors.

21. Perspectives on Contemporary Statistics, *David C. Hoaglin and David S. Moore*, Editors.

22. Heeding the Call for Change: Suggestions for Curricular Action, *Lynn A. Steen*, Editor.

24. Symbolic Computation in Undergraduate Mathematics Education, *Zaven A. Karian*, Editor.

25. The Concept of Function: Aspects of Epistemology and Pedagogy, *Guershon Harel and Ed Dubinsky*, Editors.

26. Statistics for the Twenty-First Century, *Florence and Sheldon Gordon*, Editors.

27. Resources for Calculus Collection, Volume 1: Learning by Discovery: A Lab Manual for Calculus, *Anita E. Solow*, Editor.

28. Resources for Calculus Collection, Volume 2: Calculus Problems for a New Century, *Robert Fraga*, Editor.

29. Resources for Calculus Collection, Volume 3: Applications of Calculus, *Philip Straffin*, Editor.

30. Resources for Calculus Collection, Volume 4: Problems for Student Investigation, *Michael B. Jackson and John R. Ramsay*, Editors.

31. Resources for Calculus Collection, Volume 5: Readings for Calculus, *Underwood Dudley*, Editor.

32. Essays in Humanistic Mathematics, *Alvin White*, Editor.

33. Research Issues in Undergraduate Mathematics Learning: Preliminary Analyses and Results, *James J. Kaput and Ed Dubinsky*, Editors.

34. In Eves' Circles, *Joby Milo Anthony*, Editor.

35. You're the Professor, What Next? Ideas and Resources for Preparing College Teachers, *The Committee on Preparation for College Teaching, Bettye Anne Case*, Editor.

36. Preparing for a New Calculus: Conference Proceedings, *Anita E. Solow*, Editor.

37. A Practical Guide to Cooperative Learning in Collegiate Mathematics, *Nancy L. Hagelgans, Barbara E. Reynolds, SDS, Keith Schwingendorf, Draga Vidakovic, Ed Dubinsky, Mazen Shahin, G. Joseph Wimbish, Jr.*

MAA Service Center
P. O. Box 91112
Washington, DC 20090-1112
800-331-1622   fax: 301-206-9789

Foreword

This volume, compiled by the editors on behalf of the Linear Algebra Curriculum Study Group, is for instructors and students of linear algebra as well as all those interested in the ideas of elementary linear algebra. We have noticed, through attendance at special sessions organized at the Joint Annual Meetings and through talks given at other conferences and universities, that there is broad and sustained interest in the content of undergraduate linear algebra courses.

Since this course became a centerpiece of the mathematics curriculum, beginning around 1960, new topics and new treatments have gradually reshaped it, with noticeably greater evolution than in calculus courses. In addition, current courses are often taught by those not trained in the subject or by those who learned linear algebra in a course rather different from the present one. In this setting, it is not surprising that there is considerable interest in the context and subtleties of ideas in the linear algebra course and in a perspective based upon what lies just beyond. With this in mind, we have selected 74 items and an array of problems, some previously published and some submitted in response to our request for such items (*College Mathematics Journal* **23** (1992), 299–303). We hope that these will provide a useful background and alternative techniques for instructors, sources of enrichment projects and extended problems for teachers and students, impetus for further textbook evolution to writers, and the enjoyment of discovery to others.

The Linear Algebra Curriculum Study Group (LACSG) began with a special session, at the January 1990 Joint Annual Meetings, focusing upon the elementary linear algebra course. This session was organized by Duane Porter, following upon an NSF-sponsored Rocky Mountain Mathematics Consortium Lecture Series given by Charles Johnson at the University of Wyoming; David Carlson and David Lay were panel members for that session. With NSF encouragement and support, these four organized a five-day workshop held at the College of William and Mary in August, 1990. The goal was to initiate substantial and sustained national interest in improving the undergraduate linear algebra curriculum. The workshop panel was broadly based, both geographically and with regard to the nature of institutions represented. In addition, consultants from client disciplines described the role of linear algebra in their areas and suggested ways in which the curriculum could be improved from their perspective.

Preliminary versions of LACSG recommendations were completed at this workshop and widely circulated for comment. After receiving many comments and with the benefit of much discussion, a version was published in 1993 (*College Mathematics Journal* **24** (1993), 41–46). This was followed by a companion volume to this one in 1997 (*Resources for Teaching Linear*

Algebra, MAA Notes No. 42, Washington, D.C., 1997, 297 pages). Work of the LACSG has continued with the organization of multiple special sessions at each of the Joint Annual Meetings from 1990 through 1998. With sustained strong attendance at these sessions, acknowledged influence on newer textbooks, discussion in letters to the AMS Notices, and the ATLAST workshops, the general goal of the LACSG is being met.

Though a few items in this volume are several pages, we have generally chosen short, crisp items that we feel contain an interesting idea. Previously published items have been screened from several hundred we reviewed from the past 50+ years, most from the *American Mathematical Monthly* and the *College Mathematics Journal*. New (previously unpublished) items were selected from about 100 responses to our call for contributions. Generally, we have chosen items that relate in some way to the first course, might evolve into the first course, or are just beyond it. However, second courses are an important recommendation of the LACSG, and some items are, perhaps, only appropriate at this level. Typically, we have avoided items for which both the topic and treatment are well established in current textbooks. For example, there has been dramatic improvement in the last few decades in the use of row operations and reduced echelon form to elucidate or make calculations related to basic issues in linear algebra. But, many of these have quickly found their way into textbooks and become well established, so that we have not included discussion of them. Also, because of the ATLAST volume, we have not concentrated upon items that emphasize the use of technology in transmitting elementary ideas, though this is quite important. We do not claim that each item is a "gem" in every respect, but something intrigued us about each one. Thus, we feel that each has something to offer and, also, that every reader will find something of interest.

Based upon what we found, the volume is organized into ten topical "parts." The parts and the items within each part are in no particular order, except that we have tried to juxtapose items that are closely related. Many items do relate to parts other than the one we chose. The introduction to each part provides a bit of background or emphasizes important issues about constituent items. Because of the number of items reprinted without editing, we have not adopted a common notation. Each item should be regarded as a stand-alone piece with its own notation or utilizing fairly standard notation.

Each item is attributed at the end of the item, typically in one of three ways. If it is a reprinted item, the author, affiliation at the time, and original journal reference are given. If it is an original contribution, the author and affiliation are given. In a few cases, the editors chose to author a discussion they felt important, and such items are simply attributed to "the Editors."

We would like to thank, first of all, the many colleagues and friends who contributed new items to this volume, as well as the authors of items we have chosen to reprint here. We also are grateful for the many more contributions we were unable to use due to limitations of space or fit with the final themes of the volume. We give special thanks to the National Science Foundation for its support of the activities of the Linear Algebra Curriculum Study Group (USE 89-50086, USE 90-53422, USE 91-53284), including part of the preparation of this volume, and for the support of other linear algebra education initiatives. The Mathematical Association of America receives equal thanks, not only for the publication of this volume and the previous LACSG volume (*Resources for Teaching Linear Algebra*, 1997), but also for its sponsorship of the well-attended LACSG special sessions at the January Joint Annual Meetings (along with AMS) since 1990 and for the many permissions to reprint items from MAA publications (the *American Mathematical Monthly* and the *College Mathematics Journal*) used in this volume. We thank the many attendees of these sessions for their encouragement and interest, and we thank the Rocky Mountain Mathematics Consortium and the University of Wyoming for hosting Charles Johnson's initiating lecture series and the College of William and Mary for hosting the 1990 founding workshop.

Finally, we are most grateful to our several colleagues who assisted with the preparation of this volume: John Alsup, while a PhD. student at the University of Wyoming; Joshua Porter, while an undergraduate student at the University of Wyoming; Chris Boner, while an REU student at the College of William and Mary; Fuzhen Zhang, while a visitor at the College of William and Mary; Jeanette Reisenburg for her many hours of excellent typing in Laramie; and the several reviewers for their encouragement and helpful suggestions.

Lastly, we are very pleased with the opportunity to prepare this volume. In spite of more than 125 years of collective experience in teaching linear algebra and thinking about the elementary course, we all learned a great deal from many hours of reading and then chatting about the items. Indeed, we each feel that we learn new subtleties each time we undertake to describe elementary linear algebra to a new cohort of students. We hope that all instructors will find similar value in this volume and will continue to reflect upon and enhance their understanding of ideas that underlie and relate to the course.

The Editors:

Charles R. Johnson, The College of William and Mary
David Carlson, San Diego State University
David C. Lay, University of Maryland
A. Duane Porter, University of Wyoming

CONTENTS

PART 1
Partitioned Matrix Multiplication

Introduction

The simple act of viewing a matrix in partitioned form can lead to remarkable insights. Even more remarkable is the broad utility of partitioned matrix multiplication for proofs and other insights. The first example is the variety of views of matrix multiplication itself (including the three recommended by LACSG for the first course and now appearing in several textbooks). These views are described and elaborated in the first item of this part.

More importantly, skill at general partitioned multiplication (including that of specially constructed matrices) has become the stock-in-trade of theoreticians, appliers of linear algebra, and computational linear algebraists alike. We believe, strongly, that it is also one of the best mechanisms (ranking with the row-reduced echelon form) for understanding ideas of elementary linear algebra, so that students should become aware of it early. This section includes a small selection of the very many examples that led us to this conclusion.

Modern Views of Matrix Multiplication

The classical definition of matrix product AB is an array whose (i,j)-entry is the inner product of row i of A with column j of B. Today, however, our understanding of linear algebra is greatly enhanced by viewing matrix multiplication in several different but equivalent ways.

The column definition of Ax. Let A be an $m \times n$ matrix, with columns $a_1, \ldots, a_n$, and let x be a (column) vector in $\mathbb{R}^n$, with entries $x_1, \ldots, x_n$. Then Ax is a linear combination of the columns of A, using the corresponding entries of x as the weights in the linear combination:

$$Ax = (a_1 \cdots a_n) \begin{bmatrix} x_1 \\ \vdots \\ x_n \end{bmatrix} = a_1 x_1 + \cdots + a_n x_n. \tag{1}$$

Traditionally, scalars are written to the left of vectors, and one usually writes $Ax = x_1 a_1 + \cdots + x_n a_n$.

The column definition of Ax ties the concept of linear dependence (of vectors $a_1, \ldots, a_n$) directly to the equation $Ax = 0$, and it connects the question of whether a vector b is in the subspace spanned by $a_1, \ldots, a_n$ to the question of whether the equation $Ax = b$ has a solution. Also, the single subscript notation in (1) simplifies many proofs, such as the proof that the mapping $x \to Ax$ is linear.

The column definition of AB. A definition of the matrix product AB can be given in a natural way by considering the composition of the linear mappings $x \to Bx$ and $y \to Ay$. Denote the columns of B by $b_1, \ldots, b_p$ and consider x in $\mathbb{R}^p$. Writing

$$Bx = b_1 x_1 + \cdots + b_p x_p$$

and using the linearity of the mapping $y \to Ay$, we have

$$A(Bx) = (Ab_1)x_1 + \cdots + (Ab_p)x_p.$$

Since the vector $A(Bx)$ is expressed here as a linear combination of vectors, we can use the definition (1) to write

$$A(Bx) = [Ab_1 \cdots Ab_p]x.$$

The matrix whose columns are $Ab_1, \ldots, Ab_p$ transforms x into $A(Bx)$; we denote this matrix by AB, so that

$$A(Bx) = (AB)x$$

5

for all x. Thus, matrix multiplication corresponds to composition of linear transformations when the product AB is defined by

$$AB = A[b_1 \cdots Ab_p] = [Ab_2 \cdots Ab_p]. \tag{2}$$

This view of AB makes possible, for example, simple proofs that the column space of AB is contained in the column space of A and that

$$\operatorname{rank} AB \leq \min(\operatorname{rank} A, \operatorname{rank} B).$$

The row-column rule for AB. Denote the (i, j)-entries of A, B, and AB by a_{ij}, b_{ij}, and $(AB)_{ij}$, respectively. Then definitions (1) and (2) lead to the classical "row-column rule" for computing AB:

$$(AB)_{ij} = a_{i1}b_{1j} + \cdots + a_{in}b_{nj}. \tag{3}$$

This formula holds for each i and j because the jth column of AB is Ab_j, the entries $b_{1j}, \ldots, b_{nj}$ are the weights for a linear combination of the columns of A, and the ith entries in these columns of A are $a_{i1}, \ldots, a_{in}$. The row-column rule has the advantages that it is efficient for hand computation and it can be given directly to students without a discussion of (column) vectors. Such a definition, however, is usually accompanied by an apology for giving an unnatural and unmotivated definition. (The definition, of course, can be motivated as was done historically by considering the effect of substituting one system of equations into a second set of equations and observing the form of the coefficients in the resulting new system. This approach usually requires a low-dimensional example, to avoid a nightmare of double subscripts.) The row-column rule is needed when studying orthogonality and proving that the null space of A is orthogonal to the row space of B.

Rows of AB. The row-column rule is also useful for showing that each row of the product AB is the result of right-multiplying the corresponding row of A by the matrix B. If $\operatorname{row}_i(A)$ denotes the ith row of A, then

$$\operatorname{row}_i(AB) = \operatorname{row}_i(A) \cdot B. \tag{4}$$

Because most linear algebra courses today focus on column vectors, property (4) is less useful than the column property (1). Property (4) can be used, however, to explain why a row operation on a matrix B can be produced via left-multiplication by an elementary matrix E, where E is created by performing the same row operation on the identity matrix.

The column-row expansion of AB. Another view of AB describes the product as a sum of arrays instead of an array of sums. Denoting the jth column of A by $\operatorname{col}_j(A)$, we have

$$AB = [\operatorname{col}_1(A) \cdots \operatorname{col}_n(A)] \begin{bmatrix} \operatorname{row}_1(B) \\ \vdots \\ \operatorname{row}_n(B) \end{bmatrix} = \sum_{k=1}^{n} \operatorname{col}_k(A) \cdot \operatorname{row}_k(B). \tag{5}$$

To verify (5), observe that for each row index i and column index j, the (i, j)-entry in $\operatorname{col}_k(A)\operatorname{row}_k(B)$ is the product of a_{ik} from $\operatorname{col}_k(A)$ and b_{kj} from $\operatorname{row}_k(B)$. Hence the (i, j)-entry in the sum (5) is

$$\underset{(k=1)}{a_{i1}b_{1j}} + \underset{(k=2)}{a_{i2}b_{2j}} + \cdots + \underset{(k=n)}{a_{in}b_{nj}}.$$

This sum is also the (i,j)-entry in AB, by the row-column rule. View (5) is also called an *outer-product expansion* of AB because each term is a matrix of rank at most 1.

The expansion (5) is the key ingredient in understanding the spectral decomposition of a symmetric matrix, $A = PDP^T$, in which $P = [u_1 \cdots u_n]$ and $D = \text{diag}(\lambda_1, \ldots, \lambda_n)$. First, note that $PD = [\lambda_1 u_1 \cdots \lambda_n u_n]$ because P acts on each column of D, and thus the jth column of PD is, by (1), a linear combination of the columns of P using weights that are all zero except for λ_j as the jth weight. Hence, using (5),

$$A = (PD)P^T = [\lambda_1 u_1 \cdots \lambda_n u_n] \begin{bmatrix} u_1^T \\ \vdots \\ u_n^T \end{bmatrix} = \lambda_1 u_1 u_1^T + \cdots + \lambda_n u_n u_n^T.$$

Multiplying partitioned matrices. If A is an $m \times n$ matrix and B is a $q \times p$ matrix, then the product AB is defined if and only if $n = q$, and then AB is $m \times p$. Suppose that $n = q$, and that A and B are partitioned matrices. Suppose that the rows of A, columns of A, rows of B, and columns of B are partitioned according to $\alpha = (\alpha_1 = (1, \ldots, m_1), \alpha_2 = (m_1 + 1, \ldots, m_2), \ldots)$, $\beta = (\beta_1, \beta_2, \ldots)$, $\delta = (\delta_1, \delta_2, \ldots)$, and $\gamma = (\gamma_1, \gamma_2, \ldots)$ respectively. The product AB can be calculated using these partitionings if and only if $\beta = \delta$, and in this case the rows of AB and the columns of AB are partitioned according to α and γ, respectively.

Of possible partitions of any $<t> = (1, \ldots, t)$, the coarsest is of course $\mu = (\mu_1 = <t>)$ and the finest is $\nu = (\nu_1 = (1), \ldots, \nu_t = (t))$. Suppose that A is an $m \times n$ matrix and B is an $n \times p$ matrix. There are eight possible choices of $(\alpha; \beta = \delta; \gamma)$ involving coarsest μ and finest ν partitions. The descriptions of AB for each of these follows. They give in various forms the views of matrix multiplication discussed above.

$\alpha; \quad \beta(= \delta); \quad \gamma \qquad$ **View of matrix multiplication** :

$\mu; \quad \mu; \quad \mu \qquad$ This just says $AB = A \cdot B$.

$\mu; \quad \mu; \quad \nu \qquad A \cdot (\text{col}_1(B), \ldots, \text{col}_q(B)) = (A \cdot \text{col}_1(B), \ldots, A \cdot \text{col}_q(B))$ which is (2)

$\mu; \quad \nu; \quad \mu \qquad$ The column-row expansion of AB:

$$[\text{col}_1(A) \cdots \text{col}_n(A)] \begin{bmatrix} \text{row}_1(B) \\ \vdots \\ \text{row}_n(B) \end{bmatrix} = \sum_{k=1}^{n} \text{col}_k(A) \cdot \text{row}_k(B) \qquad \text{which is (5)}$$

$$\nu; \quad \mu; \quad \mu \qquad \begin{pmatrix} \text{row}_1(A) \\ \vdots \\ \text{row}_m(A) \end{pmatrix} \cdot B = \begin{pmatrix} \text{row}_1(A) \cdot B \\ \vdots \\ \text{row}_m(A) \cdot B \end{pmatrix} \qquad \text{which is (4)}$$

$\mu; \quad \nu; \quad \nu \qquad$ From (1), each column of AB is a linear combination of the columns of A, with coefficients from that column of B:

$$(\operatorname{col}_1(A), \dots, \operatorname{col}_n(A)) \cdot \begin{pmatrix} b_{11} & \cdots & b_{1p} \\ \vdots & b_{ij} & \vdots \\ b_{n1} & \cdots & b_{np} \end{pmatrix} = \left(\cdots, \sum_{k=1}^{n} b_{kj} \operatorname{col}_k(A), \cdots \right)$$

$\nu \quad \mu \quad \nu$ The usual row-column rule for AB:

$$\begin{pmatrix} \operatorname{row}_1(A) \\ \vdots \\ \operatorname{row}_m(A) \end{pmatrix} \cdot (\operatorname{col}_1(B), \dots, \operatorname{col}_p(B)) = \begin{pmatrix} \operatorname{row}_1(A) \cdot \operatorname{col}_1(B) & \cdots & \operatorname{row}_1(A) \cdot \operatorname{col}_p(B) \\ \vdots & \vdots & \vdots \\ \operatorname{row}_m(A) \cdot \operatorname{col}_1(B) & \cdots & \operatorname{row}_m(A) \cdot \operatorname{col}_p(B) \end{pmatrix}$$

$\nu \quad \nu \quad \mu$ Analogous to (1), each row of AB is a linear combination of the columns of B, with coefficients from that row of A:

$$\begin{pmatrix} a_{11} & \cdots & a_{1n} \\ \vdots & a_{ij} & \vdots \\ a_{m1} & \cdots & a_{mn} \end{pmatrix} \cdot \begin{pmatrix} \operatorname{row}_1(B) \\ \vdots \\ \operatorname{row}_n(B) \end{pmatrix} = \begin{pmatrix} \vdots \\ \sum_{k=1}^{n} a_{ik} \cdot \operatorname{row}_k(B) \\ \vdots \end{pmatrix}$$

$\nu \quad \nu \quad \nu$ This is another version of the row-column rule.

An instructor could give and justify the alternative views presented above, and use them to lead into a general discussion of the multiplication of partitioned matrices. Or, an instructor could begin with that general discussion, and ask students to find the coarsest and finest examples, that is, to discover these new views themselves.

The Editors
Contributed

The Associativity of Matrix Multiplication

Among the often-used algebraic properties that matrix multiplication *does* satisfy, associativity, $A(BC) = (AB)C$, is perhaps the most subtle for students, and the property, therefore, merits explanation in the elementary course. If matrices are viewed as representing linear transformations, then associativity may be viewed as a simple conceptual consequence of the associativity of the composition of functions, but this is probably not helpful for many students, especially as the discussion of associativity typically comes quite early in the course. On the other hand, the most common proof via manipulation of sums with its flurry of subscripts may not be much more helpful. A disguised version of the functional composition view may be both conceptual and concrete enough to be informative to beginning students. It is also based upon an appreciation of some of the fundamental views of the mechanics of matrix multiplication.

First, note two important facts about matrix multiplication.

(a) If B is a matrix and $\mathbf{x}$ is a vector for which $B\mathbf{x}$ is defined, then the vector $B\mathbf{x}$ is a linear combination of the *columns* of B, with coefficients the entries of $\mathbf{x}$. That is,

$$B\mathbf{x} = x_1\mathbf{b}_1 + x_2\mathbf{b}_2 + \cdots + x_n\mathbf{b}_n,$$

in which the n columns of B are

$$\mathbf{b}_1, \ldots, \mathbf{b}_n :$$
$$B = \begin{bmatrix} \mathbf{b}_1 & \mathbf{b}_2 & \cdots & \mathbf{b}_n \end{bmatrix},$$

and

$$\mathbf{x} = \begin{bmatrix} x_1 \\ x_2 \\ \vdots \\ x_n \end{bmatrix}.$$

(b) If A and B are matrices for which AB is defined, then the ith column of AB is A times the ith column of B, i.e.

$$AB = \begin{bmatrix} A\mathbf{b}_1 & A\mathbf{b}_2 & \cdots & A\mathbf{b}_n \end{bmatrix}.$$

Both these facts, among others, should be stressed shortly after the definition of matrix multiplication. Now, let $\mathbf{x}$ be an arbitrary column vector. By (a) and the fact that matrix

9

multiplication acts linearly, we have

$$A(B\mathbf{x}) = A(x_1\mathbf{b}_1 + \cdots + x_n\mathbf{b}_n) = x_1 A\mathbf{b}_1 + \cdots + x_n A\mathbf{b}_n$$

$$= \begin{bmatrix} A\mathbf{b}_1 & A\mathbf{b}_2 & \cdots & A\mathbf{b}_n \end{bmatrix} \mathbf{x} = (AB)\mathbf{x},$$

the last step being observation (b). Now, if we let $\mathbf{x}$ be an arbitrary column of $C = \begin{bmatrix} \mathbf{c}_1 & \cdots & \mathbf{c}_p \end{bmatrix}$, we have

$$A(B\mathbf{c}_i) = (AB)\mathbf{c}_i$$

or again using (b)

$$A(BC) = (AB)C$$

which is associativity.

The Editors
College Mathematics Journal **23** (1992), 299–303

Relationships Between AB and BA

Let A be an $m \times n$ matrix and B be an $n \times m$ matrix with $m \leq n$. In the elementary course we stress that AB need not equal BA. In fact, they need not even be the same dimensions if $m \neq n$! However, it is important to realize that AB and BA are not independent and, in fact, have much in common. In particular the two matrices have the same eigenvalues, counting multiplicities, with BA having an additional $n - m$ eigenvalues equal to 0. This fact is seldom mentioned in the first course and may not be known to some instructors. With the following proof, perhaps it could be included in the first course. First notice that the $(m + n) \times (m + n)$ partitioned matrices

$$\begin{bmatrix} AB & 0 \\ B & 0 \end{bmatrix} \quad \text{and} \quad \begin{bmatrix} 0 & 0 \\ B & BA \end{bmatrix}$$

are similar to each other via the partitioned block calculation:

$$\begin{bmatrix} AB & 0 \\ B & 0 \end{bmatrix} \begin{bmatrix} I_m & A \\ 0 & I_n \end{bmatrix} = \begin{bmatrix} AB & ABA \\ B & BA \end{bmatrix} = \begin{bmatrix} I_m & A \\ 0 & I_n \end{bmatrix} \begin{bmatrix} 0 & 0 \\ B & BA \end{bmatrix}.$$

Since

$$\begin{bmatrix} I_m & A \\ 0 & I_n \end{bmatrix}$$

is nonsingular, it provides the similarity. Because

$$\begin{bmatrix} AB & 0 \\ B & 0 \end{bmatrix}$$

is block triangular, its eigenvalues are those of the two diagonal blocks, AB and the $n \times n$ $\mathbf{0}$ matrix. Similarly, the eigenvalues of

$$\begin{bmatrix} 0 & 0 \\ B & BA \end{bmatrix}$$

are the eigenvalues of BA, together with m 0's. Because the two partitioned matrices are similar and similar matrices have the same eigenvalues, AB and BA must have the same nonzero eigenvalues (counting multiplicities) and the additional $n - m$ eigenvalues of BA must all be 0, verifying the assertion made earlier. With only a little more effort it may be seen that the Jordan structure associated with each *non*zero eigenvalue is the same in AB and BA. The Jordan structure of the eigenvalue 0 can be different, even when $m = n$, though there is a close relation [1].

   Aside from simplicity, the above proof reinforces the use of similarity and the important tool of manipulating partitioned matrices.

Reference

1. C.R. Johnson and E. Schreiner, The Relationship Between AB and BA, *American Mathematical Monthly* **103**(1996), 578–582.

The Editors
College Mathematics Journal **23** (1992), 299–303

The Characteristic Polynomial of a Partitioned Matrix

A few years ago, I needed to know about eigenvalues of real matrices of the form

$$M = \begin{pmatrix} 0 & B \\ C & 0 \end{pmatrix}$$

where B, C are both $n \times n$. There was good reason to suspect that the characteristic polynomial $p_M(x)$ of such a matrix is an *even* function, and so it turns out to be. Perhaps more interesting is that the *same* simple argument that establishes the evenness of $p_M(x)$ also shows that BC and CB have the same eigenvalues (at least when B and C are both square).

Begin with the following lemma, which although known, is already attractive.

Lemma. Let $X = \begin{pmatrix} A & B \\ C & D \end{pmatrix} \in M_{2n}(\mathbb{C})$ where $A, B, C, D \in M_n(\mathbb{C})$. Then

(a) D and C commute $\Longrightarrow \det X = \det(AD - BC)$;

(b) A and C commute $\Longrightarrow \det X = \det(AD - CB)$.

[There are two similar results if either D and B or A and B commute.]

I will prove only (a), since the other variants are proved similarly. First take care of the "generic" case where D is invertible.

$$\det \begin{pmatrix} A & B \\ C & D \end{pmatrix} = \det \begin{pmatrix} A & B \\ C & D \end{pmatrix} \cdot \det \begin{pmatrix} I & 0 \\ -D^{-1}C & I \end{pmatrix} = \det \begin{pmatrix} A - BD^{-1}C & B \\ 0 & D \end{pmatrix}$$

$$= \det(A - BD^{-1}C) \cdot \det D$$
$$= \det(AD - BD^{-1}CD)$$
$$= \det(AD - BC).$$

Now, even if D is not invertible, $D + \epsilon I$ will be invertible for all sufficiently small ϵ. So using the above result we have

$$\det \begin{pmatrix} A & B \\ C & D + \epsilon I \end{pmatrix} = \det \Big(A(D + \epsilon I) - BC \Big).$$

The continuity of det gives the desired result, letting $\epsilon \to 0$. $\qquad \square$

I first saw a result of this type in Hoffman and Kunze's *Linear Algebra*, second edition, section 6.5 on Simultaneous Triangularization and Diagonalization (p. 208, #4). There they supply the stronger-than-necessary hypothesis that A, B, C, D should *all* commute.

From this lemma we immediately get the following proposition.

Proposition. For $M = \begin{pmatrix} 0 & B \\ C & 0 \end{pmatrix}$ where $B, C \in M_n(\mathbb{C})$ we have

(a) $p_M(x) = p_{BC}(x^2)$ and

(b) $p_M(x) = p_{CB}(x^2)$

showing twice over that $p_M(x)$ is indeed even.

Proof.

$$p_M(x) = \det(xI - M) = \det\begin{pmatrix} xI & -B \\ -C & xI \end{pmatrix} = \det(x^2 I - BC) \quad \text{[by Lemma (a)]}$$
$$= \det(x^2 I - CB) \quad \text{[by Lemma (b)]}. \quad \square$$

And now for free we have $p_{BC}(x^2) = p_{CB}(x^2)$, hence $p_{BC}(x) = p_{CB}(x)$. Thus BC and CB have the same eigenvalues, including multiplicity.

Notes

- Observing that $\begin{pmatrix} 0 & B \\ C & 0 \end{pmatrix}$ is actually *similar* to $\begin{pmatrix} 0 & C \\ B & 0 \end{pmatrix}$ via $\begin{pmatrix} 0 & I \\ I & 0 \end{pmatrix}$, and using *just* Proposition (a), we get an alternative proof that $p_{BC} = p_{CB}$.

- The argument showing that $p_{BC} = p_{CB}$ as a whole can be simplified and made even more elementary by shifting the continuity from the lemma to the proposition. Prove the lemma *only* for D is invertible; then in the proposition we would have $p_M(x) = \det(x^2 I - BC)$ for all $x \neq 0$. Both sides being polynomials then implies equality for all x.

- Let $N = \begin{pmatrix} A & 0 \\ 0 & D \end{pmatrix}$ and $M = \begin{pmatrix} 0 & B \\ C & 0 \end{pmatrix}$ with $A, B, C, D \in M_n(\mathbb{C})$. The contrast between the identities $p_N(x) = p_A(x) \cdot p_D(x)$ and $p_M(x) = p_{BC}(x^2) \neq p_B(x) \cdot p_C(x)$ is rather striking.

Editors' Note: When $B = C^*$ in M we have the familiar, but important, fact that the eigenvalues of M are $+$ and $-$ the singular values of C, leading to many important facts.

D. Steven Mackey
Western Michigan University
Contributed

The Cauchy-Binet Formula

In the following, entries in the indicated matrices are from an arbitrary field. The classical formula attributed to Binet and to Cauchy relates minors of a product to certain minors of the factors. It is seldom proven in textbooks and can be very useful in particular situations. For example, a matrix is called totally positive if *all* of its minors are positive. Cauchy-Binet implies that the totally positive matrices are closed under matrix multiplication.

Definition: If A is an $m \times n$ matrix, $\alpha \subseteq \{1, 2, \dots, m\}$ and $\beta \subseteq \{1, 2, \dots, n\}$, then $A[\alpha, \beta]$ denotes the $|\alpha| \times |\beta|$ submatrix of A with rows lying in α and columns lying in β.

Now suppose that A is an $m \times n$ matrix, B is an $n \times p$ matrix, and let $C = AB$. Let $N = \{1, 2, \dots, n\}$. Then if $\alpha \subseteq \{1, 2, \dots, m\}$ and $\beta \subseteq \{1, 2, \dots, p\}$, it follows from the definition of matrix multiplication that

$$C[\alpha, \beta] = A[\alpha, N]B[N, \beta].$$

Theorem (Cauchey-Binet Formula). *Let A be an $m \times n$ matrix, B be an $n \times p$ matrix and let $C = AB$. Given $\alpha \subseteq \{1, 2, \dots, m\}$ and $\beta \subseteq \{1, 2, \dots, p\}$ with $|\alpha| = |\beta| = r$, then*

$$\det C[\alpha, \beta] = \sum_{\substack{K \subseteq N \\ |K| = r}} \det A[\alpha, K] \det B[K, \beta] \tag{1}$$

in which the sum is over all r-element subsets K of N.

Remark. If $r = 1$ and $\alpha = \{i\}, \beta = \{j\}$, then (1) reduces to

$$c_{ij} = \sum_{k=1}^{n} a_{ik} b_{kj}$$

which is just the definition of matrix multiplication.

If we relabel $C[\alpha, \beta]$ as C, $A[\alpha, N]$ as A, and $B[N, \beta]$ as B, then $C = AB$ and (1) becomes, with $M = \{1, 2, \dots, m\}$,

$$\det(AB) = \sum_{\substack{K \subseteq N \\ |K| = m}} \det A[M, K] \det B[K, M] \tag{2}$$

15

where A is an $m \times n$ matrix, B is an $n \times m$ matrix, and $m \leq n$. (If $n = m$, (2) reduces to $\det(AB) = \det A \cdot \det B$, the multiplicative law for the determinant.) So to prove the Cauchy-Binet formula, it suffices to prove (2). If A is a square matrix, $E_k(A)$ denotes the sum of all $k \times k$ principal minors of A.

Lemma. Let A be an $m \times n$ matrix and B an $n \times m$ matrix, $m \leq n$. Then

$$E_m(BA) = \sum_{\substack{K \subseteq N \\ |K|=m}} \det A[M, K] \det B[K, M].$$

Proof. $E_m(BA) = \sum_{\substack{K \subseteq N \\ |K|=m}} \det(BA)[K, K]$

Since $(BA)[K, K] = B[K, M]A[M, K]$,

$$\det(BA)[K, K] = \det B[K, M] \cdot \det A[M, K]$$

and the result follows.

The following proof of (2) was communicated by Abraham Berman, from page 54 of *Matrix Analysis* by R. Horn and C. Johnson:

$$\begin{bmatrix} I & A \\ 0 & I \end{bmatrix}^{-1} \begin{bmatrix} AB & 0 \\ B & 0 \end{bmatrix} \begin{bmatrix} I & A \\ 0 & I \end{bmatrix} = \begin{bmatrix} 0 & 0 \\ B & BA \end{bmatrix},$$

so

$$C_1 = \begin{bmatrix} AB & 0 \\ B & 0 \end{bmatrix} \quad \text{and} \quad C_2 = \begin{bmatrix} 0 & 0 \\ B & BA \end{bmatrix}$$

are similar matrices.

Since similar matrices have the same characteristic polynomial, and since, in general, $E_k(X)$ is the coefficient of λ^{n-k} in the characteristic polynomial of the $n \times n$ matrix X,

$$E_m(C_1) = E_m(C_2)$$

which reduces to

$$E_m(AB) = E_m(BA).$$

But AB is $m \times m$, so

$$E_m(AB) = \det(AB).$$

Then by the lemma,

$$\det(AB) = \sum_{\substack{K \subseteq N \\ |K|=m}} \det A[M, K] \det B[K, M]$$

which is (2). □

Wayne Barrett
Brigham Young University
Contributed

Cramer's Rule

Though it should not be advocated as a practical means for solving systems of linear equations $A\mathbf{x} = \mathbf{b}$, in which A is $n \times n$, $\mathbf{x}$ is $n \times 1$, and $\mathbf{b}$ is $n \times 1$, Cramer's rule is a historically fundamental idea of linear algebra that is useful for the analytic representation of the sensitivity of a component of the solution $\mathbf{x}$ to the data $A, \mathbf{b}$. It is frequently taught in the elementary course, either without proof or with an algebra-intensive computational proof that is more drudgery than illuminating. Actually it can be viewed as a corollary to the multiplicativity of the determinant function, while again reinforcing important mechanics of matrix multiplication. To see this, adopt the notation $A \overset{i}{\leftarrow} \mathbf{b}$ for the $n \times n$ matrix that results from replacing the ith column of A with the column vector $\mathbf{b}$ and rewrite the system

$$A\mathbf{x} = \mathbf{b}$$

as

$$A\left[I \overset{i}{\leftarrow} \mathbf{x}\right] = A \overset{i}{\leftarrow} \mathbf{b}.$$

That the latter is an expanded version of the former again simply makes use of observation (b) from page 9. Taking determinants of both sides and using the multiplicativity of the determinant function then gives

$$(\det A)\left(\det\left[I \overset{i}{\leftarrow} \mathbf{x}\right]\right) = \det\left[A \overset{i}{\leftarrow} \mathbf{b}\right].$$

But $\det\left[I \overset{i}{\leftarrow} \mathbf{x}\right] = x_i$ by Laplace expansion, so that if $\det A \neq 0$, we have

$$x_i = \frac{\det\left[A \overset{i}{\leftarrow} \mathbf{b}\right]}{\det A},$$

which is Cramer's rule. This proof also gives practice in the important skill of exploiting the structure of sparse matrices, in this case $I \overset{i}{\leftarrow} \mathbf{x}$.

The Editors
College Mathematics Journal **23** (1992), 299–303

The Multiplicativity of the Determinant

The idea for the following proof was submitted by William Watkins who credited his idea to a discussion of 3×3 determinants in [1]. Two facts are assumed:

(i) $\det EC = \det C$ if E is a product of type III (row replacement) elementary matrices.

(ii) The determinant of a 2×2 block triangular matrix is the product of the determinants of the diagonal blocks [2].

Let A and B be $n \times n$ matrices. Then

$$(\det A)(\det B) = \det \begin{bmatrix} A & 0 \\ -I & B \end{bmatrix} = \det \begin{bmatrix} I & A \\ 0 & I \end{bmatrix} \cdot \det \begin{bmatrix} A & 0 \\ -I & B \end{bmatrix}$$

$$= \det \begin{bmatrix} I & A \\ 0 & I \end{bmatrix} \begin{bmatrix} A & 0 \\ -I & B \end{bmatrix}$$

because $\begin{bmatrix} I & A \\ 0 & I \end{bmatrix}$ is a product of type III elementary matrices. Continuing,

$$(\det A)(\det B) = \det \begin{bmatrix} 0 & AB \\ -I & B \end{bmatrix}$$

$$= (-1)^n \det \begin{bmatrix} AB & 0 \\ B & -I \end{bmatrix} \qquad n \text{ column interchanges}$$

$$= (-1)^n (\det AB) \cdot \det(-I)$$

$$= \det AB.$$

References

1. Burnside, William and Arthur Panton, *The Theory of Equations, vol. II*, Dover, New York, 1960, 29–30. Reprinted from 7th ed., Longmens, Green and Company, 1928.

2. Lay, David, *Linear Algebra and Its Applications*, 2nd ed., Addison Wesley, Reading, MA, 1997, p. 207.

The Editors

William E. Watkins
California State University, Northridge
Contributed and Edited

PART 2
Determinants

Introduction

Leibnitz formalized the determinant in 1693, and there are strong Japanese claims to knowledge of it in 1683 with likely earlier precursors. But after a longer history than other linear algebraic concepts, the determinant has become controversial! Since it is seldom numerically efficient to compute a solution to a problem by calculating a determinant, and since it is difficult to define the determinant so cleanly in infinite dimensions, the notion of determinant is sometimes downplayed by computational linear algebraists and operator theorists.

Nonetheless, it is difficult to imagine a more fundamental single scalar to associate with a square matrix, and experience has long demonstrated its ability to contribute unique insights on both theoretical and applied levels. The structural beauty of determinants is not in question, and, as students find relative ease with their concrete nature, they remain a core topic of elementary linear algebra. However, it is difficult for the first course to include more than the tip of the iceberg of analytic power determinants hold.

Muir's volumes on the history of determinants contain an immense amount of classical information, both tedious and fundamental, so that it is difficult to assemble many "fresh" items about such an historical topic. We have included here a number of items that are less well-known and can enrich the elementary course. Because of its centrality, there are remarkably many ways to calculate a determinant. The definition, Laplace expansion, elementary operations, and Schur complements are familiar. Less familiar and amusing ways are discussed in the pieces by Barrett and by Fuller and Logan, and these are only a sample. The fundamental theorem of multiplicativity of the determinant lies in contrast to the fact that the determinant of a sum has no simple relation to the sum of the determinants. Several aspects of what is true are discussed in the first two papers of this part. Each of the other pieces contains an elementary idea that may intrigue teachers as well as students.

The Determinant of a Sum of Matrices

If $A = (a_{ij})$ and $B = (b_{ij})$ are $n \times n$ matrices such that $a_{ij} = b_{ij}$ for $i \neq s$ and $C = (c_{ij})$ is an $n \times n$ matrix such that $c_{ij} = a_{ij}$ for $i \neq s$ and $c_{sj} = a_{sj} + b_{sj}$ for $j = 1, 2, \ldots, n$ then $|C| = |A| + |B|$. This property of determinants leads some students to the incorrect generalization that $|A + B| = |A| + |B|$ for any matrices A and B.

The formula for $|A+B|$ is actually a bit more involved; since, going back to the definition of determinants, we find that $|A + B| = \sum (a_{1i_1} + b_{1i_1}) \cdot (a_{2i_2} + b_{2i_2}) \cdots (a_{ni_n} + b_{ni_n})$ where $i_1, i_2, \ldots, i_n$ range over all permutations of $1, 2, \ldots, n$. If, in evaluating this sum of products, we choose first all the a_{ij} and sum over the permutations of $1, 2, \ldots, n$, we obtain the value $|A|$. If we choose the b_{ij}, we obtain $|B|$. If we make some other definite choice of a_{ij} or b_{ij} from each parenthesis, and sum over the permutations of $1, 2, \ldots, n$, we obtain the determinant of a matrix whose kth row is the kth row of A or B depending upon whether we chose a_{ij} or b_{ij} from the kth parenthesis.

There are 2^n possible ways of making the above choices; therefore, we see that we may evaluate $|A + B|$ by constructing the 2^n matrices each of which is obtained by choosing for its kth row the kth row of A or B and then taking the sum of the determinants of these matrices.

This result may be extended to $|A_1 + A_2 + \cdots + A_t|$. We find in a similar manner that $|A_1 + A_2 + \cdots + A_t|$ is the sum of the determinants of t^n matrices each of which is obtained by choosing for its kth row one of the A_i.

The student should note that if $a_1, a_2, \ldots, a_t$ are elements of some non-commutative ring, there is a strong analogy between the expressions for $|A_1+A_2+\cdots+A_t|$ and $(a_1+a_2+\cdots+a_t)^n$ in that $(a_1 + a_2 + \cdots + a_t)^n$ is the sum of t^n products each of which is obtained by choosing for its kth factor one of the a_i.

L.M. Weiner
DePaul University
American Mathematical Monthly **63** (1956), 273

Determinants of Sums

A Determinant Formula. In an issue of the *College Mathematics Journal* [Evaluating "uniformly filled" determinants, *CMJ* 19, 1988, pp. 343–345], S.M. Goberstein exhibits a formula for computing the determinant of a matrix V obtained from a matrix U by adding the scalar v to every entry of U. The author then evaluates several determinants of this type and mentions that freshmen mathematics majors in the Soviet Union have used the method for several decades.

The method that Goberstein exhibits for determinant of this type is a very special case of an interesting formula for the determinant of the sum of any two matrices. This formula for $\det(A+B)$ can be easily derived from the Laplace expansion theorem, and consequently is readily accessible to students in an elementary linear algebra course. As we shall see, the formula can also be used to directly obtain important results about the characteristic polynomial, and about relationships between the characteristic roots and subdeterminants of A. The formula in question is one of those "folk" results whose precise origins are difficult to trace. It appears in [Marvin Marcus, *Finite Dimensional Multilinear Algebra, Part II*, Marcel Dekker Inc., New York, 1975, pp. 162–163] with a related result about the permanent. [Ed: If $A = (a_{ij})$, per $(A) \equiv \sum \prod_{i=1}^{n} a_{i\sigma(i)}$, in which the sum is over all permutations σ. Thus, the permanent is the determinant without the $-$ signs.] In fact, the permanent version of the formula provides a simple proof of the classical formula that counts the number of derangements of an n element set [Herbert J. Ryser, *Combinatorial Mathematics*, The Mathematical Association of American, Carus Mathematical Monograph #14, 1963, pp. 22–28]. However, I first learned the determinant version from Professor Emilie Haynsworth more than twenty years ago. The formula is

$$\det(A + B) = \sum_r \sum_{\alpha,\beta} (-1)^{s(\alpha)+s(\beta)} \det\left(A\left[\alpha|\beta\right]\right) \det\left(B(\alpha|\beta)\right). \tag{1}$$

In (1): A and B are n-square matrices; the outer sum on r is over the integers $0, \ldots, n$, for a particular r, the inner sum is over all strictly increasing integer sequences α and β of length r chosen from $1, \ldots, n$; $A[\alpha|\beta]$ (square brackets) is the r-square submatrix of A lying in rows α and columns β; $B(\alpha|\beta)$ is the $(n-r)$-square submatrix of B lying in rows complementary to α and columns complementary to β; and $s(\alpha)$ is the sum of the integers in α. Of course, when $r = 0$ the summand is taken to mean $\det(B)$ and when $r = n$, it is $\det(A)$.

The proof of (1) is a very simple consequence of the linearity of the determinant in each row of the matrix, and the standard Laplace expansion theorem. Here are the details of the

29

argument. Write

$$\det(A + B) = \det\left(A_{(1)} + B_{(1)}, \ldots, A_{(n)} + B_{(n)}\right)^T \tag{2}$$

where $A_{(i)}$ denotes the ith row of A. The right side of (2) formally acts just like a product of the "binomials" $A_{(i)} + B_{(i)}, i = 1, \ldots, n$: this is the meaning of det being linear in the rows. Thus for each r chosen from $0, \ldots, n$, the right side of (2) contributes a sum over all α of length r of terms of the form

$$\det\left(B_{(\alpha'_1)}, \ldots, A_{(\alpha_1)}, \ldots, A_{(\alpha_r)}, \ldots, B_{(\alpha'_{n-r})}\right) \tag{3}$$

in which $A_{(\alpha_t)}$ occupies row position α_t in (3), $t = 1, \ldots, r$ and $B_{(\alpha'_t)}$ occupies row position α'_t, $t = 1, \ldots, n - r$. The sequence α' is strictly increasing and complementary to α in $1, \ldots, n$. Let X_α denote the n-square matrix in (3), i.e., $A_{(\alpha_t)}$ is row α_t of $X_\alpha, t = 1, \ldots, r$ and $B_{(\alpha'_t)}$ is row α'_t of $X_\alpha, t = 1, \ldots, n - r$. (In the cases $r = 0$ and $r = n, X_\alpha$ is B and A, respectively.) In terms of X_α we have

$$\det(A + B) = \sum_r \sum_\alpha \det(X_\alpha). \tag{4}$$

Use the Laplace expansion on rows numbered α in X_α to obtain

$$\det(X_\alpha) = (-1)^{s(\alpha)} \sum_\beta (-1)^{s(\beta)} \det\left(X_\alpha[\alpha|\beta]\right) \det\left(X_\alpha\right)(\alpha|\beta). \tag{5}$$

But according to the definition of X_α,

$$X_\alpha[\alpha|\beta] = A[\alpha|\beta], \tag{6}$$

and

$$\begin{aligned} X_\alpha(\alpha|\beta) &= X_\alpha[\alpha'|\beta'] \\ &= B[\alpha'|\beta'] \\ &= B(\alpha|\beta). \end{aligned} \tag{7}$$

Substitute (6) and (7) in (5) and then replace $\det(X_\alpha)$ in (4) by (5). The result is formula (1).

Some Examples. The examples in Goberstein's article are all of the form $A + B$ in which B has rank 1. Any rank 1 matrix is of the form $B = uv^T$ where u and v are nonzero $n \times 1$ matrices. Every subdeterminant of uv^T of size 2 or more is 0, so that the only summands that survive on the right side of the formula (1) are those corresponding to $r = n$ and $r = n - 1$. For $r = n$ the single summand is $\det(A)$; for $r = n - 1$ the interior sum may be rewritten as

$$\sum_{i,j=1}^n (-1)^{i+j} \det\left(A(i|j)\right) u_i v_j. \tag{8}$$

The i, j entry of $\mathrm{adj}(A)$, the adjugate of A (sometimes called the adjoint of A), is $(-1)^{i+j} \det A(j|i)$ and hence (8) can be written as $v^T \mathrm{adj}(A)u$. Thus we have the rather neat formula

$$\det(A + uv^T) = \det(A) + v^T \mathrm{adj}(A)u. \tag{9}$$

The first matrix considered in Goberstein's article is

$$\text{diag}(1 - n, 2 - n, 3 - n, \dots, 0) + nJ \tag{10}$$

in which J is the n-square matrix all of whose entries are 1. Let A be the diagonal matrix in (10) and let $u = ne, v = e$ where e is the $n \times 1$ matrix all of whose entries are 1. Obviously $\det(A) = 0$ and the only term that survives in (8) corresponds to $i = j = n$, namely

$$n \cdot (1 - n)(2 - n) \cdots 1 = (-1)^{n-1} n!.$$

The second and third matrices considered in Goberstein's article are

$$-P + J \tag{11}$$

and the $2n$-square matrix

$$M + J. \tag{12}$$

In (11), P is the matrix whose only nonzero entries are 1's on the sinister diagonal, i.e., $P_{i,n-i+1} = 1, \dots, n$; the matrix M in (12) is a direct sum of n copies of the 2-square matrix

$$\begin{bmatrix} 0 & 1 \\ 1 & 0 \end{bmatrix},$$

and J is the $2n$-square matrix of 1's. In (11) note that $\det(P) = (-1)^p$ where $p = [n/2]$, i.e., the largest integer in $n/2$. Since $P^2 = I_n$ we have $P^{-1} = P$,

$$\begin{aligned} \text{adj}(P) \ &= \det(P) \cdot P^{-1} \\ &= (-1)^p P. \end{aligned}$$

Then

$$\begin{aligned} \text{adj}(-P) \ &= \det(-P)(-P)^{-1} \\ &= (-1)^{n+1} \det(P) \cdot P \\ &= (-1)^{n+1}(-1)^p P \\ &= (-1)^{n+p+1} P, \end{aligned}$$

so that (9) becomes

$$\begin{aligned} \det(-P + J) \ &= \det(-P) + e^T \text{adj}(-P)e \\ &= (-1)^{n+p} + (-1)^{n+p+1} e^T P e \\ &= (-1)^{n+p}(1 - n). \end{aligned}$$

It is easy to check that $n + p$ and $n(n + 1)/2$ have the same parity and thus

$$\det(-P + J) = (-1)^{n(n+1)/2}(1 - n). \tag{13}$$

The value of $\det(M + J)$ is equally simple to compute. Observe that M also satisfies $M^{-1} = M$, and that $\det(M) = (-1)^n$. Thus, as was the case with the matrix P,

$$\text{adj}(M) = (-1)^n M,$$

and again taking $u = v = e$, (9) specializes to

$$
\begin{aligned}
\det(M + J) &= \det(M) + (-1)^n e^T M e \\
&= (-1)^n + (-1)^n (2n) \\
&= (-1)^n (1 + 2n).
\end{aligned}
\tag{14}
$$

The arguments used so far point to a unified formula incorporating (13) and (14). Let A be an arbitrary n-square real orthogonal matrix. Then

$$
\begin{aligned}
\mathrm{adj}(A) &= \det(A) A^{-1} \\
&= \det(A) A^T.
\end{aligned}
$$

If $u = ce$ and $v = e$ then $A + uv^T = A + cJ$ and (9) becomes

$$
\begin{aligned}
\det(A + cJ) &= \det(A) + \det(A) c e^T A^T e \\
&= \det(A)(1 + ca)
\end{aligned}
\tag{15}
$$

where a is the sum of the entries in A. A referee points out that the formula (15) holds for matrices that satisfy $e^T A e = e^T A^{-1} e$. The proof is identical.

One of the troika of excellent referees assigned to this paper suggests several additional examples on which to apply (9) or some variant of it. The first of these is the matrix

$$
P + B \tag{16}
$$

where B is the n-square rank 1 matrix whose kth row is

$$
B_{(k)} = k[1\,2\,3\ldots n], \qquad k = 1, \ldots, n,
$$

and P is the matrix in (11). Note that $B = uv^T$ with

$$
u = v - [1\,2\,3\ldots n]^T.
$$

Then (9) becomes

$$
\begin{aligned}
\det(P + B) &= \det(P) + (-1)^p u^T P u \\
&= (-1)^p \left(1 + \sum_{k=1}^{n} k(n - k + 1) \right).
\end{aligned}
$$

The summation in this last formula is quickly evaluated as

$$
n(n+1)(n+2)/6
$$

and thus

$$
\det(P + B) = (-1)^p (1 + n(n+1)(n+2)/6), \qquad p = n \operatorname{div} 2. \tag{17}
$$

As another example we will evaluate the function

$$
f(x) = \det(I_n + B) \tag{18}
$$

where B is the n-square rank 1 matrix whose kth row is

$$
B_{(k)} = x^k [1\, x\, \cdots\, x^{n-1}], \qquad k = 1, \ldots, n
$$

(x is indeterminate). Note that

$$B = xuu^T$$

where

$$u = [1 \; x \; x^2 \; \cdots \; x^{n-1}]^T.$$

Again applying (9) we have

$$
\begin{aligned}
f(x) &= 1 + xu^T u \\
&= 1 + x \sum_{k=0}^{n-1} x^{2k} \\
&= 1 + \sum_{k=0}^{n-1} x^{2k+1}.
\end{aligned}
$$

The general idea of how to use (9) should be clear, so we leave a final example as an exercise for the reader: evaluate

$$f(x) = \det(D + xJ)$$

where $D = \operatorname{diag}(1, 2, \ldots, n)$. The answer is

$$f(x) = n! \left(1 + x \sum_{k=1}^{n} \frac{1}{k} \right).$$

The Characteristic Polynomial. The formula (1) can be directly applied to the characteristic matrix $\lambda I_n - B$ by replacing A by λI_n and B by $-B$. Since $\det(I_n[\alpha|\beta])$ is 0 for $\alpha \neq \beta$ and is 1 for $\alpha = \beta$, we have

$$
\begin{aligned}
\det(\lambda I_n - B) &= \sum_{r=0}^{n} \sum_{\alpha} \lambda^r \det(-B(\alpha|\alpha)) \\
&= \sum_{r=0}^{n} \lambda^r (-1)^{n-r} b_{n-r}
\end{aligned}
\tag{19}
$$

where b_{n-r} is the sum of all $(n-r)$-square principal subdeterminants of B. If $\lambda_1, \ldots, \lambda_n$ are characteristics roots of B then

$$
\begin{aligned}
\det(\lambda I_n - B) &= \prod_{i-1}^{n} (\lambda - \lambda_i) \\
&= \sum_{r=0}^{n} \lambda^r (-1)^{n-r} e_{n-r}
\end{aligned}
\tag{20}
$$

where e_k is the kth elementary symmetric polynomial in $\lambda_1, \ldots, \lambda_n$. Matching coefficients in (19) and (20) we have

$$e_k = b_k, \qquad k = 1, \ldots, n.$$

Of course, $k = 1$ and $k = n$ are the familiar

$$\operatorname{tr}(B) = \sum_{i=1}^{n} \lambda_i,$$

and

$$\det(B) = \prod_{i=1}^{n} \lambda_i.$$

Acknowledgement. This work was supported by the Air Force Office of Scientific Research under grant AFOSR-88-0175.

Marvin Marcus
University of California, Santa Barbara
College Mathematics Journal **21** (1990), 130–135

An Application of Determinants

A simple and elegant application of the theory of determinants for the beginning student is the following criterion of Sylvester, a well-known theorem of algebraic lore: let K be a field and $f(x) = a_m x^m + \cdots + a_1 x + a_0$, $g(x) = b_n x^n + \cdots + b_1 x + b_0$, where $a_m \neq 0 \neq b_n$, be two polynomials in $K[x]$; then $f(x)$ and $g(x)$ have a nonconstant factor in $K[x]$ if and only if the determinant of the following $(m+n) \times (m+n)$ matrix A is zero:

$$
A = \begin{bmatrix}
a_m & a_{m-1} & \cdots a_1 & a_0 & 0 & 0 & \cdots 0 \\
0 & a_m & \cdots a_2 & a_1 & a_0 & 0 & \cdots 0 \\
\cdot & \cdot & \cdots & \cdot & \cdot & \cdot & \cdots \\
0 & 0 & \cdots & & & & \cdots a_0 \\
b_n & b_{n-1} & \cdots & & b_0 & 0 & \cdots 0 \\
0 & b_n & \cdots & & b_1 & b_0 & \cdots 0 \\
\cdot & \cdot & \cdots & \cdot & \cdot & \cdot & \cdots \\
0 & 0 & \cdots & & & & \cdots b_0
\end{bmatrix}.
$$

We present a simple proof of this theorem which requires only knowledge of the fact that the determinant of the product of two matrices is the product of the determinants; no use of the theory of linear equations is needed. Set

$$
B = \begin{bmatrix}
x^{n+m-1} & 0 & 0 & \cdot & \cdot & \cdot & 0 \\
x^{n+m-2} & 1 & 0 & \cdot & \cdot & \cdot & 0 \\
x^{n+m-3} & 0 & 1 & \cdot & \cdot & \cdot & 0 \\
\cdot & & & \cdot & \cdot & \cdot & \cdot \\
x & & 0 & 0 & \cdot & \cdot 1 & 0 \\
1 & & 0 & 0 & \cdot & \cdot & \cdot 1
\end{bmatrix}.
$$

Then $|B| = x^{n+m-1}$ and

$$
|AB| = |A| x^{n+m-1} = \begin{vmatrix}
x^{n-1} f(x) & a_{m-1} & a_{m-2} & \cdots 0 \\
x^{n-2} f(x) & a_m & a_{m-1} & \cdots 0 \\
\cdot & \cdot & \cdot & \cdots \\
f(x) & \cdot & \cdot & \cdots a_0 \\
x^{m-1} g(x) & b_{n-1} & b_{n-2} & \cdots 0 \\
\cdot & \cdot & \cdot & \cdots \\
g(x) & \cdot & \cdot & \cdots b_0
\end{vmatrix} = f(x) h(x) + g(x) k(x),
$$

where $h(x)$ and $k(x)$ are polynomials in $K[x]$ of degree at most $n-1$ and $m-1$, respectively, calculated by expanding $|AB|$ by the first column.

35

If $f(x)$ and $g(x)$ have a nonconstant factor $r(x)$, then

$$|A|x^{n+m-1} = f(x)h(x) + g(x)k(x) = r(x)q(x),$$

for some polynomial $q(x)$. If $q(x) = 0$, then clearly $|A| = 0$. If $q(x) \neq 0$, then $r(x)$ is a multiple of some power of x. But since $r(x)$ is a factor of both $f(x)$ and $g(x)$, both a_0 and b_0 must be zero, whence the last column of A consists of zeros and again $|A| = 0$.

Conversely, suppose $|A| = 0$. Then $f(x)h(x) = -g(x)k(x)$. Factoring both sides of this equality into irreducible factors over K we must obtain the same factors, and hence all factors of $f(x)$ must divide either $g(x)$ or $k(x)$. But since $k(x)$ is of at most degree $m - 1$, not all factors of $f(x)$ can divide $k(x)$, hence $f(x)$ and $g(x)$ have a common factor.

Helen Skala
University of Massachusetts, Boston
American Mathematical Monthly **78** (1971), 889–890

Cramer's Rule via Selective Annihilation

For an instructive classroom session that employs important concepts in linear algebra, begin by solving

$$a_{11}x_1 + a_{12}x_2 = b_1 \qquad (a_{11}a_{22} - a_{12}a_{21} \neq 0) \qquad (1)$$
$$a_{21}x_1 + a_{22}x_2 = b_2$$

using Cramer's well-known rule:

$$x_1 = \frac{\begin{vmatrix} b_1 & a_{12} \\ b_2 & a_{22} \end{vmatrix}}{\begin{vmatrix} a_{11} & a_{12} \\ a_{21} & a_{22} \end{vmatrix}}.$$

This result may be easily rederived by writing (1) in vector form as

$$x_1\mathbf{a_1} + x_2\mathbf{a_2} = \mathbf{b}, \qquad (2)$$

where

$$\mathbf{a_1} = \begin{bmatrix} a_{11} & a_{21} \end{bmatrix}^T, \quad \mathbf{a_2} = \begin{bmatrix} a_{12} & a_{22} \end{bmatrix}^T, \quad \mathbf{b} = \begin{bmatrix} b_1 & b_2 \end{bmatrix}^T.$$

Since the vector $\mathbf{v} = \begin{bmatrix} a_{22} & -a_{12} \end{bmatrix}^T$ is orthogonal to $\mathbf{a_2}$, taking its dot product with both sides of (2) leads to

$$x_1\mathbf{a_1} \cdot \mathbf{v} = \mathbf{b} \cdot \mathbf{v},$$

or

$$x_1 = \frac{\mathbf{b} \cdot \mathbf{v}}{\mathbf{a_1} \cdot \mathbf{v}} = \frac{b_1 a_{22} - b_2 a_{12}}{a_{11}a_{22} - a_{21}a_{12}} = \frac{\begin{vmatrix} b_1 & a_{12} \\ b_2 & a_{22} \end{vmatrix}}{\begin{vmatrix} a_{11} & a_{12} \\ a_{21} & a_{22} \end{vmatrix}}.$$

Thus, by annihilating one term in the vector equation we are able to determine one of the unknowns.

This idea may be generalized to three dimensions in a natural way. Given the vector equation

$$x_1\mathbf{a_1} + x_2\mathbf{a_2} + x_3\mathbf{a_3} = \mathbf{b}$$

for noncoplanar vectors $\mathbf{a_1}, \mathbf{a_2}, \mathbf{a_3}$, we wish to take the dot product of both sides with a vector $\mathbf{v}$ that will annihilate both $\mathbf{a_2}$ and $\mathbf{a_3}$. The cross product is the obvious choice, and leads to

$$x_1\mathbf{a_1} \cdot \mathbf{a_2} \times \mathbf{a_3} = \mathbf{b} \cdot \mathbf{a_2} \times \mathbf{a_3}.$$

Thus

$$x_1 = \frac{\mathbf{b} \cdot \mathbf{a_2} \times \mathbf{a_3}}{\mathbf{a_1} \cdot \mathbf{a_2} \times \mathbf{a_3}}$$

When the triple products are expressed as determinants, the Cramer's Rule formula for x is obtained.

The idea of eliminating all but one of the vector terms by an appropriate dot product may be extended to n equations in n unknowns. Consider the equation

$$x_1\mathbf{a_1} + x_2\mathbf{a_2} + \cdots + x_n\mathbf{a_n} = \mathbf{b}, \tag{3}$$

where $\{\mathbf{a_i} = (a_{i1}, a_{i2}, \ldots, a_{in}) : 1 \le i \le n\}$ are assumed to be independent vectors in $\mathbb{R}^n$ To generate a vector $\mathbf{v}$ orthogonal to $\mathbf{a_2}, \ldots, \mathbf{a_n}$, for example, form the formal array

$$\begin{bmatrix} \mathbf{e_1} & \mathbf{e_2} & \cdots & \mathbf{e_n} \\ a_{21} & a_{22} & \cdots & a_{2n} \\ \vdots & \vdots & & \vdots \\ a_{n1} & a_{n2} & \cdots & a_{nn} \end{bmatrix},$$

where $\mathbf{e_i}$ is the ith standard basis vector, and expand as a determinant along the first row. The resulting linear combination of $\mathbf{e_1}, \ldots, \mathbf{e_n}$ is the desired vector $\mathbf{v}$. Note that, for any vector $\mathbf{c} = [c_1, \ldots, c_n]^T$, the value of $\mathbf{v} \cdot \mathbf{c}$ is given by

$$\mathbf{v} \cdot \mathbf{c} = \det \begin{bmatrix} c_1 & c_2 & \cdots & c_n \\ a_{21} & a_{22} & \cdots & a_{2n} \\ \vdots & \vdots & & \vdots \\ a_{n1} & a_{n2} & \cdots & a_{nn} \end{bmatrix},$$

verifying that $\mathbf{v} \cdot \mathbf{a_i} = 0$ for $i = 2, \ldots, n$. Alternatively, observe that the explicit formulation of $\mathbf{v}$ is not necessary. We simply require a way to compute $\mathbf{v} \cdot \mathbf{c}$ for any vector $\mathbf{c}$. Accordingly, for $1 \le i \le m$, define a linear functional L_i by the formula

$$L_i(\mathbf{c}) = \det(\mathbf{a_1}, \ldots, \mathbf{a_{i-1}}, \mathbf{c}, \mathbf{a_{i+1}}, \ldots, \mathbf{a_n})$$

for each vector $\mathbf{c}$. Then $L_i(\mathbf{a_j}) = 0$ for $j \ne i$, and applying L_i to both sides of (3) gives

$$x_i L_i(\mathbf{a_i}) = L_i(\mathbf{b}).$$

This is equivalent to Cramer's Rule.

Dan Kalman
The Aerospace Corporation
College Mathematics Journal **18** (1987), 136–137

The Multiplication of Determinants

There are perhaps half a dozen different proofs given in the literature to the classical theorem asserting that the determinant of the product of two square matrices is equal to the product of the two determinants of the individual matrix factors. There is, of course, always the vague question as to how much variation in two proofs should entail a feeling that the two proofs are really essentially different. But nobody is likely to gainsay that the three proofs given in the standard texts of Bôcher, Kowalewski, and Birkhoff and MacLane are quite different from each other.

The purpose of this note is to present still another proof which I have used in the classroom and which has been particularly suitable from the point of view of conciseness and also from the point of view of what properties of determinants and matrices it is convenient to develop beforehand. Until recently I have never seen in the existing literature anything like it. But in the preparation of this note a more thorough search has revealed the idea in a book by G. Doster entitled *Éléments de la théorie des determinants*, [pp. 65–71]. The method is there incompletely developed; it is applied explicitly only to three-rowed matrices and even so misses some points of rigor.

Our proof, as distinguished from most other proofs of the same theorem, uses analysis to obtain a purely algebraic result. Indeed the fundamental theorem of algebra (which despite its name is a theorem of analysis rather than of algebra) is fundamental to our method. We also use the further result from analysis, deduced from the implicit function theorem, to the effect that the roots of an algebraic equation of the nth degree are continuous functions of the coefficients, at least as long as the n roots are distinct.

Theorem 1. *If P and Q are both $n \times n$ matrices, if P is non-singular and if Q has distinct eigenvalues, none of which are zero, then the two equations*

$$(1) \qquad\qquad \det(xI - Q) = 0$$

and

$$(2) \qquad\qquad \det(xP - PQ) = 0,$$

which are both equations of the n-th degree in x, have the same roots all of which are simple and none of which is zero.

Proof. x is a root of (1) if and only if there exists an n-vector $v \neq 0$ such that

$$(3) \qquad\qquad (xI - Q)v = 0$$

and it is a root of (2) if and only if there exists an n-vector $v \neq 0$ such that

$$(4) \qquad\qquad (xP - PQ)v = 0.$$

39

But (3) is readily transformed into (4) by multiplying on the left by P, and (4) likewise is transformed by (3) by multiplying on the left by the inverse of P, which is assumed to exist. It follows that every root of (1) is a root of (2) and every root of (2) is a root of (1). Since the roots of (1) are the n distinct eigenvalues of Q, none zero, it is clear that the nth degree equation (2), as well as (1), must have just n simple distinct roots, none zero.

Theorem 2. *If P and Q are as in the previous theorem, then*

$$(5) \qquad \det(PQ) = (\det P)(\det Q).$$

Proof. Since, by Theorem 1, the equations (1) and (2) have simple identical roots, their coefficients must be proportional. Now the coefficients of x^n in these two equations are 1 and $\det P$, respectively, while their constant terms are $(-1)^n \det Q$ and $(-1)^n \det(PQ)$ respectively. $\det(PQ) \neq 0$, since the case of a zero root was excluded from (1) and hence also from (2). We are therefore fully justified in writing $1/\det P = \det Q/\det(PQ)$, from which (5) follows at once.

Theorem 3. *If A and B are arbitrary $n \times n$ matrices*

$$\det(AB) = (\det A)(\det B).$$

Proof. Let C and D be $n \times n$ matrices such that C is non-singular, while D has distinct eigenvalues, none zero. Define the matrices $P(t)$ and $Q(t)$ as follows:

$$P(t) = tA + (1-t)C, \; Q(t) = tB + (1-t)D,$$

so that

$$(6) \qquad P(1) = A, \qquad Q(1) = B$$

and

$$(7) \qquad P(0) = C, \qquad Q(0) = D.$$

Also let

$$(8) \qquad F(t) = \det(P(t)Q(t)) - (\det P(t))(\det Q(t)).$$

$F(t)$ is evidently a polynomial in t, which, if not identically zero, is of degree not exceeding $2n$. Because of (7), we know that $P(0)$ is non-singular and that $Q(0)$ has distinct eigenvalues, each different from zero. By continuity we know that the same is true for $P(t)$ and $Q(t)$ as long as $|t|$ is sufficiently small. Hence by Theorem 2 and (8) we know that $F(t) = 0$ for all t whose absolute values are sufficiently small. Since $F(t)$ is known to be a polynomial, we must therefore have $F(t) \equiv 0$. In particular $F(1) = 0$, and the theorem then follows at once from (6) and (8).

With Theorem 3 we have reached our main objective: but it will probably occur to the reader that Theorem 3 is merely a special case of the following.

Theorem 4. *Let p_k be the coefficient of x^k in the polynomial $\det(xI - B)$ and let q_k be the coefficient of x^k in the polynomial $\det(xA - AB)$, where A and B are arbitrary $n \times n$ matrices and where $k = 0, 1, 2, \cdots, n$. Then $p_k q_j = p_j q_k$.*

We leave the proof as an exercise to the reader and merely remark that Theorem 3 is the special case of Theorem 4 where $k = 0$ and $j = n$.

Theorem 3 also admits an entirely different generalization, the so-called Binet-Cauchy formula relating to the product of certain types of rectangular matrices [4]. There appears to be no way by which the methods of the present note can be modified so as to yield the Binet-Cauchy formula directly. However, even though Theorem 3 is generally deduced as a consequence of the Binet-Cauchy formula in most of the texts where the latter is discussed, it is also possible to deduce the Binet-Cauchy formula as a consequence of Theorem 3.

References

1. Garrett Birkhoff and Saunders MacLane, *A Survey of Modern Algebra*, Macmillan, New York, 1941, pp. 288–289.
2. Maxime Bôcher, *Introduction to Higher Algebra*, Macmillan, New York, 1915, pp. 26–27.
3. G. Dostor, *Éléments de la théorie des déterminants*, Gauthier-Villars et Cie, Paris, 1921, pp. 65–71.
4. F.R. Gantmacher, *The Theory of Matrices, vol. 1*, Chelsea, New York, 1960, pp. 9–10.
5. Gerhard Kowalewski, *Einführung in die Determinantentheorie*, 3rd Edition, Chelsea, New York, 1948, pp. 60–62.

D.C. Lewis
Baltimore, MD
American Mathematical Monthly **83** (1976), 268–270

A Short Proof of a Result of Determinants

In the reference below, Busch proves the following:

Theorem. *If A is an $(n-1) \times n$ matrix of integers such that the row sums are all zero, then $|AA'| = nk^2$ where k is an integer.* (A' means A transpose, and $|A|$ means determinant of A.)

The proof uses an interesting application of the Cauchy-Binet theorem, prefaced by a remark on how the unwary may be led astray if they improperly use the rules for manipulating rows of a determinant. It is possible, by partitioning A, to get a simpler proof—one which suggests generalizations and also provides the value of k. Write $A = (B, B\beta')$, where B is $(n-1) \times (n-1)$ and $\beta = (-1, -1, \cdots, -1)$ is $1 \times (n-1)$. Then

$$
\begin{aligned}
|AA'| &= \left| (B, B\beta) \begin{pmatrix} B' \\ \beta B' \end{pmatrix} \right| = |BB' + B\beta'\beta B'| = |B(I + \beta'\beta)B'| \\
&= |B|^2 |I + \beta'\beta|.
\end{aligned}
$$

Since $|I + \beta'\beta| = n$, we have $|AA'| = n|B|^2$.

Reference

1. K.A. Busch, An interesting result on almost square matrices and the Cauchy-Binet Theorem, *American Mathematical Monthly*, **69** (1962) 648– 649.

George Marsaglia
Boeing Scientific Research Laboratories
American Mathematical Monthly **72** (1965), 173

43

Dodgson's Identity

We begin with the 3×3 matrix

$$M = \begin{bmatrix} 4 & 2 & 1 \\ 2 & 3 & 2 \\ 1 & 5 & 6 \end{bmatrix}$$

and evaluate its determinant as follows. Compute the determinant of each of the four contiguous 2×2 submatrices of M and record its value in the "center" of each:

$$\begin{vmatrix} 4 & & 2 & & 1 \\ & 8 & & 1 & \\ 2 & & 3 & & 2 \\ & 7 & & 8 & \\ 1 & & 5 & & 6 \end{vmatrix}.$$

Disregarding the "edge" entries, we compute the determinant of the "new" 2×2 submatrix and divide by the center entry, yielding

$$\frac{\begin{vmatrix} 8 & 1 \\ 7 & 8 \end{vmatrix}}{3} = \frac{57}{3} = 19,$$

which is the determinant of M as can be checked by conventional methods.

More generally, we have the following.

Dodgson's identity. Let M be an $n \times n$ matrix and let A be the submatrix obtained by deleting the last row and column, B by deleting the last row and first columns, C by deleting the first row and last column, D by deleting the first row and column, and E by deleting the first and last rows and first and last columns. Let a, b, c, d, e be the respective determinants of these submatrices and assume that e is nonzero (so E is invertible). Then

$$\det M = \frac{\begin{vmatrix} a & b \\ c & d \end{vmatrix}}{e}. \tag{1}$$

A convenient mnemonic representation for (1) is

This identity may be viewed as a direct generalization of the rule for evaluating 2×2 determinants. Note that unlike the usual expansion rules for determinants such as the minor expansion and Laplace's expansion, which involve products from non-overlapping submatrices, all five of these submatrices overlap in the "big" submatrix E.

We give the following elementary proof. Write M as a partitioned matrix

$$M = \begin{bmatrix} r & v^T & s \\ w & E & x \\ t & y^T & u \end{bmatrix} .$$

Since E is invertible, v^T and y^T are each linear combinations of the rows of E, while w and x are linear combinations of the columns of E. Applying the corresponding elementary row operations to M gives the matrix

$$M' = \begin{bmatrix} r' & 0^T & s' \\ 0 & E & 0 \\ t' & 0^T & u' \end{bmatrix}$$

where $\det M' = \det M$. Furthermore, the determinants of the submatrices of M' that correspond to the submatrices A, B, C, D in M are unchanged because no row or column outside any one of these submatrices has been added to one inside. Thus, $a = \det A = \det \begin{bmatrix} r' & 0^T \\ 0 & E \end{bmatrix} = r'e$, and likewise $b = (-1)^{n-2}s'e$, $c = (-1)^{n-2}t'e$ and $d = u'e$. Upon expanding $\det M'$ we have $\det M' = (r'u' - s't')e$, which yields $\det M = (ad - bc)/e$. $\square$

Dodgson's identity can be applied sequentially to evaluate the determinant of an $n \times n$ matrix: Begin by computing the determinants of all the contiguous 2×2 submatrices, apply the identity to all of them to yield the determinants of all contiguous 3×3 submatrices, and so on. We illustrate with the following 4×4 example:

$$A = \begin{bmatrix} 1 & 3 & 2 & -1 \\ 2 & -1 & 3 & 1 \\ -1 & 2 & 1 & 1 \\ -2 & -5 & 2 & 3 \end{bmatrix} .$$

Step 1: Evaluate the contiguous 2×2 submatrices:

$$\begin{vmatrix} 1 & & 3 & & 2 & & -1 \\ & -7 & & 11 & & 5 & \\ 2 & & -1 & & 3 & & 1 \\ & 3 & & -7 & & 2 & \\ -1 & & 2 & & 1 & & 1 \\ & 9 & & 9 & & 1 & \\ -2 & & -5 & & 2 & & 3 \end{vmatrix}.$$

Eliminate the entries on the edge as they are no longer needed:

$$\begin{vmatrix} -7 & & 11 & & 5 \\ & -1 & & 3 & \\ 3 & & -7 & & 2 \\ & 2 & & 1 & \\ 9 & & 9 & & 1 \end{vmatrix}.$$

Step 2: Evaluate the contiguous 3×3 submatrices by Dodgson's identity.

$$\begin{vmatrix} -7 & & 11 & & 5 \\ & -16 & & 19 & \\ 3 & & -7 & & 2 \\ & 45 & & -25 & \\ 9 & & 9 & & 1 \end{vmatrix}.$$

For example,

$$-16 = \frac{\begin{vmatrix} -7 & 11 \\ 3 & -7 \end{vmatrix}}{-1} = \begin{vmatrix} 1 & 3 & 2 \\ 2 & -1 & 3 \\ -1 & 2 & 1 \end{vmatrix}.$$

Again eliminating the entries on the edge, we arrive at

$$\begin{vmatrix} -16 & & 19 \\ & -7 & \\ 45 & & -25 \end{vmatrix}.$$

Step 3:

$$\frac{\begin{vmatrix} -16 & 19 \\ 45 & -25 \end{vmatrix}}{-7} = 65,$$

which is the determinant of A.

We leave it as an exercise to show that this method of evaluating an $n \times n$ determinant requires approximately n^3 multiplications and divisions to carry out. Note that it is also highly parallelizable, as the computation of the determinants of each $k \times k$ submatrix in step $k - 1$ can be done independently. Nonetheless, this method cannot be regarded as a serious computational tool because the process terminates if one ever has to divide by zero before reaching the answer. This cannot in general be anticipated beforehand and there seems to be no way to effectively compensate. However, for the 3×3 and 4×4 determinants encountered in the classroom one can quickly check for a problem with zeros before beginning the computation, and, if necessary, avoid it by a judicious row/column interchange.

History. Dodgson's identity is a special case of two earlier determinantal identities, Jacobi's formula for minors of the inverse and Sylvester's identity [4]. See [1] for a survey of several classical determinantal identities. The above method for evaluating determinants was introduced by C.L. Dodgson (better know as the author Lewis Carroll) in [2]. It has no doubt been rediscovered independently several times (see e.g. [5]) and has appeared in a textbook [3]. I also discovered Dodgson's identity and his method for evaluating determinants in 1978–79 and even used his term "condensation" to describe the process. At the time I verified Dodgson's identity by means of the above proof; that the identity and procedure are credited to him I learned at a very interesting talk given by Alfred W. Hales at the annual meeting of the Mathematical Association of America in Phoenix, 1989. He presented exactly the same proof and reported on more general identities of this type and their application to the solution of a problem in combinatorics [6]. As this account is surely incomplete, I would appreciate learning from readers of other occurrences of Dodgson's identity.

References

1. R.A. Brualdi and H. Schneider, Determinantal Identities: Gauss, Schur, Cauchy, Sylvester, Krone, *Linear Algebra and Its Applications* **52/53** (1983), 769-791, and **59**(1984), 203–207.
2. C.L. Dodgson, Condensation of Determinants, *Proceedings of the Royal Society of London* **15** (1866), 150–155.
3. P.S. Dwyer, *Linear Computations*, Wiley, New York, 1951, pp. 147–148.
4. F.R. Gantmacher, *The Theory of Matrices, volume I*, Chelsea, New York, 1959, pp. 21, 32.
5. R.H. Macmillan, Contractants: A New Method for the Numerical Evaluation of Determinants, *Journal of the Royal Aeronautical Society* **59** (1955), 772–773.
6. D.P. Robbins and H. Rumsey, Jr., Determinants and Alternating Sign Matrices, *Advances in Mathematics* **62** (1986), 169–184.

Wayne Barrett
Brigham Young University
Contributed

On the Evaluation of Determinants by Chiò's Method

1. Introduction

The traditional methods for hand-calculating the determinant of an $n \times n$ matrix are based either on expansion along a row or column (expansion by minors), or on simplifying the determinant by performing elementary row operations (adding a multiple of one row to another, and so on). In addition, for second- and third-order determinants, there are well-known schemes which permit the calculation to be done quickly; but unfortunately, and especially to the consternation of students, these special schematic devices do not work for higher order determinants.

The purpose of this note is to publicize a concise, simple, alternative method for hand-calculating a determinant of arbitrary order. This method, a special case of which was originally described by Chiò [1] in 1853, owes its utility to the fact that it requires only the calculation of second-order determinants. Other treatments of Chiò's method appear in Whitaker and Robinson [2] and Muir [3], but these treatments, which give proofs based on Sylvester's method, are somewhat dated with respect to language and notation; the method does not in general find its way into more modern textbooks, and it is relatively unknown to many mathematicians. In our treatment in the following paragraphs, we have hoped to make Chiò's method accessible to the beginning student of matrix algebra; this includes a precise formulation of a generalization of the method and a proof which depends only on some elementary facts about determinants.

2. The Reduction Theorem.

Given an n-th order determinant, the procedure begins by reducing it to a single $(n-1)$st order determinant, and then reducing the $(n-1)$st order determinant to a single $(n-2)$nd order determinant, and so on until a simple second order determinant is reached. At each step, the reduction for a kth order determinant to a $(k-1)$st order determinant requires only the calculation of $(k-1)^2$ second order determinants. Consequently, the evaluation of an nth order determinant can be carried out by evaluating $(n-1)^2 + (n-2)^2 + \cdots + 3^2 + 2^2 + 1$ second-order determinants.

The basic reduction theorem can be formulated as follows:

Theorem. *Let $A = (a_{ij})$ be an $n \times n$ matrix for which $a_{nn} \neq 0$, and let $E = (e_{ij})$ be the $(n-1) \times (n-1)$ matrix defined by*

$$e_{ij} = a_{ij}a_{nn} - a_{in}a_{nj}, i, j = 1, \ldots, n-1.$$

49

Then

$$\det E = a_{nn}^{n-2} \det A.$$

We note that the elements of e_{ij} of the $(n-1) \times (n-1)$ matrix E consist of the determinants of certain two-by-two submatrices of A; that is,

$$e_{ij} = \det \begin{pmatrix} a_{ij} & a_{in} \\ a_{nj} & a_{nn} \end{pmatrix}, \quad (i, j = 1, \dots, n-1).$$

We now give the proof of the reduction theorem.

Proof: From the matrix $A = (a_{ij})$ we construct a new $n \times n$ matrix B by multiplying the first $n-1$ rows of A by the non-zero element a_{nn}. Hence

$$B = \begin{pmatrix} a_{11}a_{nn} & a_{12}a_{nn} & \cdots & a_{1n}a_{nn} \\ \vdots & \vdots & & \vdots \\ a_{n-1,1}a_{nn} & a_{n-1,2}a_{nn} & \cdots & a_{n-1,n}a_{nn} \\ a_{n1} & a_{n2} & \cdots & a_{nn} \end{pmatrix}$$

Therefore, we clearly have

$$\det B = a_{nn}^{n-1} \det A. \tag{1}$$

Now, let C be the matrix obtained from B by replacing the ith row of B by its ith row minus a_{in} times its last row; we do this for $i = 1, 2, \dots, n-1$. Then,

$$C = \begin{pmatrix} a_{11}a_{nn} - a_{1n}a_{n1} & \cdots & a_{1,n-1}a_{nn} - a_{1n}a_{n,n-1} & 0 \\ \vdots & & \vdots & \vdots \\ a_{n-1,1}a_{nn} - a_{n-1,n}a_{n1} & \cdots & a_{n-1,n-1}a_{nn} - a_{n-1,n}a_{n,n-1} & 0 \\ a_{n1} & \cdots & a_{n,n-1} & a_{nn} \end{pmatrix}$$

or

$$C = \left(\begin{array}{ccc|c} & & & 0 \\ & E & & \vdots \\ & & & 0 \\ \hline a_{n1} & \cdots & a_{n,n-1} & a_{nn} \end{array} \right)$$

where E is the desired reduced matrix. Since C was obtained from B by elementary row operations, it follows that

$$\det C = \det B \tag{2}$$

but, upon expanding C by minors along the last column, we obtain

$$\det C = a_{nn} \det E. \tag{3}$$

Equations (1),(2), and (3) imply

$$\det E = a_{nn}^{n-2} \det A,$$

and the theorem is proved.

Instead of a_{nn}, the statement of the theorem can be extended to any non-zero element a_{rs} of A. This can be accomplished by interchanging the rth and nth rows and the sth and nth columns, thereby placing the element a_{rs} in the n,n position; moreover, the two interchanges leave the sign of the original determinant unaltered. In general, it is not difficult to see that if $a_{rs} \neq 0$, then

$$\det E = a_{rs}^{n-2} \det A, \text{ where } E = (e_{ij})$$

is now given by

$$e_{ij} = (-1)^t(a_{ij}a_{is} - a_{is}a_{rj}),$$

where $t = (i - r)(j - s), t \neq 0$.

We shall call the element a_{rs} the *pivot element*. If a zero element of A is chosen as the pivot, then the determinant of E turns out to be zero, in which case the determinant of A cannot be evaluated by this method. Other treatments, as in the references, depend upon having a pivot element numerically equal to 1.

3. An Example

To illustrate this method, we evaluate the fourth order determinant

$$\begin{vmatrix} 1 & -2 & 3 & 1 \\ 4 & 2 & -1 & 0 \\ 0 & 2 & 1 & 5 \\ -3 & 3 & 1 & 2 \end{vmatrix}.$$

Choosing $a_{44} = 2$ as the pivot and applying the reduction theorem with $n = 4$, the above determinant is equal to

$$\frac{1}{2^{4-2}} \begin{vmatrix} \begin{vmatrix} 1 & 1 \\ -3 & 2 \end{vmatrix} & \begin{vmatrix} -2 & 1 \\ 3 & 2 \end{vmatrix} & \begin{vmatrix} 3 & 1 \\ 1 & 2 \end{vmatrix} \\ \begin{vmatrix} 4 & 0 \\ -3 & 2 \end{vmatrix} & \begin{vmatrix} 2 & 0 \\ 3 & 2 \end{vmatrix} & \begin{vmatrix} -1 & 0 \\ 1 & 2 \end{vmatrix} \\ \begin{vmatrix} 0 & 5 \\ -3 & 2 \end{vmatrix} & \begin{vmatrix} 2 & 5 \\ 3 & 2 \end{vmatrix} & \begin{vmatrix} 1 & 5 \\ 1 & 2 \end{vmatrix} \end{vmatrix}$$

$$= \frac{1}{4} \begin{vmatrix} 5 & -7 & 5 \\ 8 & 4 & -2 \\ 15 & -11 & -3 \end{vmatrix}.$$

In the resulting third-order determinant, we shall choose -11 as the pivot element with $n = 3$ in order to illustrate the generalization of the method. Therefore, the original determinant becomes

$$\left(\frac{1}{4}\right)\left(\frac{1}{(-11)^{3-2}}\right)$$

$$\left| \begin{array}{cc} \begin{vmatrix} 5 & -7 \\ 15 & -11 \end{vmatrix} & \begin{vmatrix} -7 & 5 \\ -11 & -3 \end{vmatrix} \\ \begin{vmatrix} 8 & 4 \\ 15 & -11 \end{vmatrix} & \begin{vmatrix} 4 & -2 \\ -11 & -3 \end{vmatrix} \end{array} \right|$$

$$= -\frac{1}{44} \begin{vmatrix} 50 & 76 \\ -148 & -34 \end{vmatrix} = -217 \; .$$

Clearly, this method is much quicker than expansion by minors, and it requires considerably less bookkeeping than any other method; it is general enough, moreover, to calculate determinants of arbitrary size, and it is exceedingly easy to master. As an alternative method for hand-calculating determinants, therefore, Chiò's method is quite effective. For numerical computations of large determinants on a computer, however, Chiò's method is not so efficient as other methods such as, for example, Gaussian elimination, because of certain difficulties with round-off errors. In addition, the method described above requires approximately $\frac{2}{3}n^3$ multiplications, whereas Gaussian elimination requires approximately $\frac{1}{3}n^3$.

References

1. F.Chiò, *Mémoire sur les Fonctions Connues sous le nom des Résultants ou des Déterminants*, Turin, 1853.
2. E.T. Whittaker and G. Robinson, *The Calculus of Observations*, Blackie and Son, London, 1924.
3. T. Muir, *A Treatise on the Theory of Determinants*, Dover (reprint), New York, 1966.

L.E. Fuller and J.D. Logan
Kansas State University
College Mathematics Journal **6** (1975), 8–10

Apropos Predetermined Determinants

In "Predetermined Determinants" [*CMJ* 16 (September 1985) 227–229], David Buchtal illustrated how determinants whose terms in arithmetic progression can be used to motivate students to study determinants and their properties. In this capsule, we extend these results to geometric progressions and to arithmetic progressions of higher order.

The value of a determinant whose terms are in geometric progression is zero, because the rows of the determinant are proportional. Thus,

$$\begin{vmatrix} 0.0625 & 0.125 & 0.25 \\ 0.5 & 1 & 2 \\ 4 & 8 & 16 \end{vmatrix} = 0 \quad \text{and} \quad \begin{vmatrix} 0.0016 & 0.008 & 0.04 \\ 0.2 & 1 & 5 \\ 25 & 125 & 625 \end{vmatrix} = 0.$$

Recall that

$$\begin{vmatrix} 1 & 2 & 3 \\ 4 & 5 & 6 \\ 7 & 8 & 9 \end{vmatrix} = 0.$$

One can also compute that

$$\begin{vmatrix} 1^2 & 2^2 & 3^2 & 4^2 \\ 5^2 & 6^2 & 7^2 & 8^2 \\ 9^2 & 10^2 & 11^2 & 12^2 \\ 13^2 & 14^2 & 15^2 & 16^2 \end{vmatrix} = 0 \quad \text{and} \quad \begin{vmatrix} 1^3 & 2^2 & 3^3 & 4^3 & 5^3 \\ 6^3 & 7^3 & 8^3 & 9^3 & 10^3 \\ 11^3 & 12^3 & 13^3 & 14^3 & 15^3 \\ 16^3 & 17^3 & 18^3 & 19^3 & 20^3 \\ 21^3 & 22^3 & 23^3 & 24^3 & 25^3 \end{vmatrix} = 0.$$

In fact, this follows from the more general cases

$$\begin{vmatrix} a^2 & (a+1)^2 & (a+2)^2 & (a+3)^2 \\ b^2 & (b+1)^2 & (b+2)^2 & (b+3)^2 \\ c^2 & (c+1)^2 & (c+2)^2 & (c+3)^2 \\ d^2 & (d+1)^2 & (d+2)^2 & (d+3)^2 \end{vmatrix} = 0$$

and

$$\begin{vmatrix} a^3 & (a+1)^3 & (a+2)^3 & (a+3)^3 & (a+4)^3 \\ b^3 & (b+1)^3 & (b+2)^3 & (b+3)^3 & (b+4)^3 \\ c^3 & (c+1)^3 & (c+2)^3 & (c+3)^3 & (c+4)^3 \\ d^3 & (d+1)^3 & (d+2)^3 & (d+3)^3 & (d+4)^3 \\ e^3 & (e+1)^3 & (e+2)^3 & (e+3)^3 & (e+4)^3 \end{vmatrix} = 0.$$

We can prove this for the 4×4 determinant of squares by proceeding as follows: reduce the second, third, and fourth columns to first-degree polynomials by adding suitable multiples

53

of the first column to each, then reduce the third and fourth columns to constants by adding a suitable multiple of the second column to each, then reduce the fourth column to zero by adding a suitable multiple of the third column to the fourth column. For the 5×5 determinant of cubes, this type of procedure reduces the fifth column to zeros.

These results can be further generalized to determinants whose entries are members of an arithmetic progression of higher order. For a given sequence $a_1, a_2, a_3, \ldots$ the sequence of differences of consecutive terms $\Delta a_1 = a_2 - a_1, \Delta a_2 = a_3 - a_2, \Delta a_3 = a_4 - a_3, \ldots$ is called the first-order difference sequence; higher-order difference sequences are formed by repeating this procedure on the preceding difference sequence. An arithmetic progression of order k is a sequence for which the kth order difference sequence is the last one that does not vanish. In other words, the kth order difference sequence is the constant sequence $d, d, d, \ldots$ with $d \neq 0$. Thus,

$$
\begin{array}{ccccc}
1 & 4 & 9 & 16 & 25\ldots \\
& 3 & 5 & 7 & 9 \quad \ldots \\
& & 2 & 2 & 2 \quad \ldots
\end{array}
\qquad \text{and} \qquad
\begin{array}{ccccc}
1 & 8 & 27 & 64 & 125\ldots \\
& 7 & 19 & 37 & 61 \quad \ldots \\
& & 12 & 18 & 24 \ldots \\
& & & 6 & 6 \qquad \ldots
\end{array}
$$

show that the consecutive squares (cubes) form an arithmetic sequence of order 2 (order 3). More generally, as is well known, the consecutive kth powers form an arithmetic sequence of order k. [See Calvin Long's "Pascal's Triangle. Difference Tables, and Arithmetic Sequences of Order N," *CMJ* 15 (September 1984) 290–298.]

Further examples are the polygonal and pyramidal numbers. The triangular numbers $\left(\frac{1}{2}\right) n(n-1)$, the pentagonal numbers $\left(\frac{1}{2}\right) n(3n-1)$, the hexagonal numbers $n(2n-1)$, etc., form arithmetic progressions of order two; the tetrahedral numbers $\binom{n}{3}$ form an arithmetic progression of order three.

Our main result is the following.

Theorem. *Let* $a_1, a_2, a_3, \ldots$ *be an arithmetic progression of order k, and let d be the constant obtained as the kth difference sequence. Then*

$$
D = \begin{vmatrix}
a_1 & a_2 & \cdots & a_n \\
a_{n+1} & a_{n+2} & \cdots & a_{2n} \\
\cdot & \cdot & \cdots & \cdot \\
\cdot & \cdot & \cdots & a_{n^2}
\end{vmatrix}
= \begin{cases}
0, & k \leq n - 2, \\
(-n)^{\binom{n}{2}} d^n, & k = n - 1.
\end{cases}
$$

To prove this, we begin by subtracting the first column from the second, the second column from the third, etc. In this way, we obtain a determinant whose second, third, $\ldots, n$th columns contain the elements of the first difference sequence. We continue by subtracting the second column from the third, the third column from the fourth, etc., to obtain a determinant whose third, fourth, $\ldots, n$th columns contain the elements of the second difference sequence. In the $(n-1)$st step, we get a determinant whose last column consists of the elements of the $(n-1)$st difference sequence. If $k \leq n - 2$ (equivalently, $n - 1 \geq k + 1$), then the last column is 0, and hence the determinant $D = 0$.

If $k \geq n - 1$, then $D \neq 0$ and one may think that the determinant is no longer "predetermined." Therefore, it may come as a surprise that in the case $k = n - 1$, the value of the determinant does not depend on the actual members of the arithmetic progression, so long as consecutive members are entered. For example:

$$
\begin{vmatrix} 1^2 & 2^2 & 3^2 \\ 4^2 & 5^2 & 6^2 \\ 7^2 & 8^2 & 9^2 \end{vmatrix} = \begin{vmatrix} 2^2 & 3^2 & 4^2 \\ 5^2 & 6^2 & 7^2 \\ 8^2 & 9^2 & 10^2 \end{vmatrix} = \begin{vmatrix} 3^2 & 4^2 & 5^2 \\ 6^2 & 7^2 & 8^2 \\ 9^2 & 10^2 & 11^2 \end{vmatrix} = \cdots = -216.
$$

This is a consequence of the following result, which may also be of interest in its own right.

Lemma. *Let $a_0, a_1, a_2, \ldots$ be an arithmetic progression of order k with $\Delta^k a_n = d$ for $n = 0, 1, 2, \ldots$. Then $a_0, a_m, a_{2m}, \ldots$ is also an arithmetic progression of order k with $\Delta^k a_{mn} = m^k d$ for $n = 0, 1, 2, \ldots$. More generally, a subsequence formed by taking every mth member, starting with an arbitrary member of the arithmetic progression of order k is also an arithmetic progression of order k.*

Proof of Lemma. An arithmetic progression of order k can be expressed as a polynomial of degree k:

$$
a_n = c_k n^k + c_{k-1} n^{k-1} + \cdots + c_0 \qquad (c_k \neq 0). \qquad\qquad (*)
$$

In more detail, $a_n = (1+\Delta)^n a_0 = \sum_{j=0}^{k} \binom{n}{j} \Delta^j a_0$ because $\Delta^{k+1} a_0 = \cdots = \Delta^n a_0 = 0$. Since the $\Delta^j a_0$ are constants and $\binom{n}{k}$ is a polynomial of degree k, we express a_n as in $(*)$. By linearity of the operator Δ, and because $\Delta^k n^h = 0$ for $h \leq k-1$, we have $\Delta^k a_n = \Delta^k c_k n^k = c_k \Delta^k n^k = d$ for $n = 0, 1, 2, \cdots$. Therefore, for

$$
a_{mn} = c_k (mn)^k + c_{k-1}(mn)^{k-1} + \cdots + c_0,
$$

we have

$$
\Delta^k a_{mn} = \Delta^k c_k m^k n^k = m^k c_k \Delta^k n^k = m^k d \qquad \text{for} \qquad n = 0, 1, 2, \ldots.
$$

The general case follows from this, because we may omit the first N members of the arithmetic progression of order k, the remainder is an arithmetic progression of the same order.

We may now conclude the proof of the Theorem as follows. In order to calculate the value of D in the case $k = n - 1$, observe that the steps prescribed in the first part of the proof have brought the original determinant to the form

$$
D = \begin{vmatrix} a_1 & \Delta a_2 & \cdots & \Delta^{n-2} a_{n-1} & \Delta^{n-1} a_n \\ a_{n+1} & \Delta a_{n+2} & \cdots & \Delta^{n-2} a_{2n-1} & \Delta^{n-1} a_{2n} \\ a_{2n+1} & \Delta a_{2n+2} & \cdots & \Delta^{n-2} a_{3n-1} & \Delta^{n-1} a_{3n} \\ a_{3n+1} & \Delta a_{3n+2} & \cdots & \Delta^{n-2} a_{4n-1} & \Delta^{n-1} a_{4n} \\ \vdots & \vdots & \cdots & \vdots & \vdots \end{vmatrix}.
$$

By the Lemma, the elements in the jth column are in arithmetic progression of order $n - j$. In particular, each element in the last column is the constant d. Moreover, the jth difference sequence of the elements in the $(n - j)$th column is constant and is equal to $n^j d$.

Therefore, we may bring the determinant to triangular form as follows. Subtract the first row from the second, the second row from the third, etc., making all the elements except the first one in the nth column equal to zero, and all the elements except the first one in the $(n-1)$st column equal to the constant value nd. Then subtract the second row from the third, the third row from the fourth, etc., making all the elements except the first two in the $(n-1)$st column equal to zero, and all the elements except the first two in the $(n-2)$nd column equal to the constant n^2d. Proceeding in this way, after the $(n-1)$st step we get

$$D = \begin{vmatrix} * & * & \cdots & * & * & d \\ * & * & \cdots & * & nd & 0 \\ * & * & \cdots & n^2d & 0 & 0 \\ \cdots\cdots\cdots\cdots\cdots\cdots\cdots\cdots \\ n^{n-1}d & 0 & \cdots & 0 & 0 & 0 \end{vmatrix} = (-1)^{\binom{n}{2}} \begin{vmatrix} n^{n-1}d & 0 & \cdots & 0 & 0 & 0 \\ \cdots\cdots\cdots\cdots\cdots\cdots\cdots\cdots \\ * & * & \cdots & n^2d & 0 & 0 \\ * & * & \cdots & * & nd & 0 \\ * & * & \cdots & * & * & d \end{vmatrix}$$

$$= (-1)^{\binom{n}{2}} n^{1+2+\cdots+n-1} d^n = (-n)^{\binom{n}{2}} d^n.$$

As the signature indicates, it takes $(n-1)+(n-2)+\cdots+2+1 = \binom{n}{2}$ switches of rows to bring the determinant to the form in which the upper right triangle consists of zeros. As the students will recall, the value of such a determinant can be calculated by taking the product of the elements in the main diagonal.

Acknowledgment. The author wishes to express his gratitude to the referees for their helpful suggestions.

Antal E. Fekete
Memorial University of Newfoundland
College Mathematics Journal **19** (1988), 254–257

PART 3
Eigenanalysis

Introduction

Eigenvectors and eigenvalues are among the most powerful of linear algebraic concepts. Eigenvectors (with eigenvalues) provide, in many instances, the right way to describe the action of a linear map. Eigenvalues, and relations between them, yield a panoply of beautiful and useful results. Eigenanalysis figures prominently in application and in functional analysis. The basic ideas are not hard: Given a linear map L on a space V, eigenvectors are the nonzero vectors on which the map acts most simply. Nevertheless, the concise definition,

$$\text{nonzero } v \text{ in } V \text{ is an eigenvector of } L \iff L(v) = tv \text{ for some scalar } t,$$

does not easily reveal its power and utility to students. (Indeed, a graduate student once asked one of the editors about writing a Master's thesis on "What are eigenvectors, anyway?")

An added difficulty for students is that eigentheory is nonlinear. (How do you find v and t so that $L(v) = tv$?) If A is an $n \times n$ matrix and $L(v) = Av$, the eigenvalues of L (and of A) are the roots of the characteristic polynomial $p(t) = \det(tI - A)$. Definitely nonlinear, and nontrivial for the numerical analyst. And, for the lower division student, the computations are more complicated than those required to row-reduce a matrix or orthogonalize a set of linearly independent vectors. They require algebra, not just arithmetic (which they all do on their calculators anyway).

Several of the articles in this chapter are intended to help the harried instructor construct reasonable examples. Others discuss numerical methods or numerical exploration for students. Others explore some of the beautiful relationships among eigenvalues, and the use of eigenanalysis to illuminate various mathematical structures. Taken together, they add a lot of meat to the bare bones of $L(v) = tv$.

Eigenvalues, Eigenvectors and Linear Matrix Equations

Understanding of linear matrix equations such as the commutativity equation

$$AX - XA = 0$$

and Lyapunov's and Sylvester's equations (e.g. [1]) hinges upon a study of the transformation from $n \times n$ matrices over a field F into itself, given by left and right multiplication by fixed matrices:

$$X \to AXB \equiv L_{A,B}(X).$$

Being a linear transformation from a vector space into itself, of course $L_{A,B}$ has eigenvalues (though this seems curious to students). Eigenanalysis of L_{AB} may be, and historically typically is (e.g. [1]), carried out by developing the theory of Kronecker products) to give an $n^2 \times n^2$ matrix representation of $L_{A,B}$. This construct, unfortunately, places the subject beyond elementary status. However, for many purposes (all but some fine points), eigenanalysis may be carried out at an elementary level and in a manner appropriate for discovery exercises or an enrichment topic in the elementary course. It also illustrates the importance of the notion of "left" (as opposed to right) eigenvector.

Suppose that A has eigenvalues $\{\lambda_i\}_{i=1}^n$ and associated right eigenvectors $\{y_i\}_{i=1}^n$, while B has eigenvalues $\{\mu_j\}_{j=1}^n$ and associated *left* $\left(z^T B = \mu_j z^T\right)$ eigenvectors $\{z_j\}_{j=1}^n$. For simplicity, we'll assume that the λ_i's are distinct and the μ_j's are distinct, though deleting this assumption poses an interesting challenge in the elementary context.

Claim: The eigenvalues of $L_{A,B}$ are $\{\lambda_i \mu_j\}_{i,j=1}^n$ and the matrices $\left\{y_i z_j^T\right\}_{i,j=1}^n$ are associated (linearly independent) eigenvectors.

Proof: Let $X = y_i z_j^T$. Then,

$$L_{A,B}(X) = A y_i z_j^T B = \lambda_i y_i \mu_j z_j^T = (\lambda_i \mu_j) y_i z_j^T = \lambda_i \mu_j X,$$

so that $\lambda_i \mu_j$ is an eigenvalue associated with the eigenvector $y_i z_j^T$. If

$$0 = \sum_{i,j} a_{ij} y_i z_j^T = Y A Z^T,$$

in which Y has the columns y_i, Z has the columns z_i, and $A = (a_{ij})$, then $A = 0$, as Y and Z^T are invertible. Thus, the set $\left\{y_i, z_j^T\right\}$ is linearly independent. The equality that results in $Y A Z^T$ exploits the Linear Algebra Curriculum Study Group's three views of matrix multiplication. [See "Modern Views of Matrix Multiplication" in Part 1 of this book.] $\square$

It is interesting that this approach shows that "eigenmatrices" of $L_{A,B}$ are generically rank 1. In general, they can be higher rank. For example, $\begin{bmatrix} 1 & 0 \\ 1 & 1 \end{bmatrix}$ is an eigenmatrix associated with the eigenvalue 0 (multiplicity 4) of $L_{\begin{bmatrix} 0 & 0 \\ 1 & 0 \end{bmatrix}, \begin{bmatrix} 0 & 0 \\ 1 & 0 \end{bmatrix}}$.

Reference

1. R. Horn and C.R. Johnson, *Topics in Matrix Analysis*, CUP, NY, 1991, Chapter 4.

Charles R. Johnson
The College of William and Mary
Contributed

Matrices with Integer Entries and Integer Eigenvalues

When eigenvalues are first covered in linear algebra, it is often convenient to give numerically simple exercises—thus it may be desirable to provide problems which have integer solutions. It is reasonably easy to find examples of 2×2 and 3×3 matrices with integer entries and integer eigenvalues, but this leads naturally to the question: exactly how can one construct, in general, an $n \times n$ matrix with such a property? The theorem below gives a complete answer, in the sense that all such matrices can be built up in the same elementary way. Remarkably, its proof is based on only three well-known results.

THEOREM. *The $n \times n$ matrix A with integer entries has integer eigenvalues if and only if it is expressible in the form*

$$A = \sum_{i=1}^{n-1} \mathbf{u}_i^T \mathbf{v}_i + k\mathbf{I}_n$$

where the $\mathbf{u}_i$ and $\mathbf{v}_i$ are row vectors with n integer components such that $\mathbf{u}_i \cdot \mathbf{v}_j = 0$ for $1 \leq i \leq j \leq n-1$, k is an integer, and I_n is the unit $n \times n$ matrix.

The eigenvalues are k, $\mathbf{u}_1 \cdot \mathbf{v}_1 + k, \dots, \mathbf{u}_{n-1} \cdot \mathbf{v}_{n-1} + k$.

Discussion and Proof. Consider the $n \times n$ matrix A with integer entries and eigenvalues. We can choose one eigenvalue, say k, and form $B = A - kI_n$. Clearly B also has integer entries and eigenvalues, one of which is 0. It is now sufficient to show that B is expressible in the form of the summation term in A above. The first step is to show:

LEMMA 1. *Let B be an $n \times n$ matrix with integer entries whose determinant is zero. Then B is expressible in the form $B = XY$ where X is $n \times (n-1)$, Y is $(n-1) \times n$, and all entries are integers.*

Proof. This is based on a well-known result (see for example Theorem 7.10 in [2]): since the determinant of B is zero, B is expressible in the form $B = M_1 D M_2$ where M_1 and M_2 are invertible and D has the form diagonal $(d_1, \dots, d_{n-1}, 0)$, all entries of M_1, M_2, and D being integers. Now $D = D_1 D$ where D_1=diagonal $(1, \dots, 1, 0)$ is $n \times n$, and $B = (M_1 D_1)(D M_2)$. But the nth column of $M_1 D_1$ and the nth row of $D M_2$ are null: delete these, and call the remainder X and Y, respectively. Then $B = XY$ as required.

Note that the converse also holds.

Two other fundamental lemmas are required: Lemma 2 may be found in the exercises of Chapter 22 in [1], attributed to Barton, while Lemma 3 is Theorem III.12 in [3]. These are:

LEMMA 2. *Let $B = XY$ where X is $n \times (n-1)$, Y is $(n-1) \times n$. Then the characteristic polynomial of B, $|B - \lambda I_n|$, is equal to the polynomial $-\lambda |YX - \lambda I_{n-1}|$.*

63

LEMMA 3. *Let C be an $m \times m$ matrix with integer entries and integer eigenvalues. Then there exists T, $m \times m$ and invertible, $|T| = \pm 1$, with integer entries such that $T^{-1}CT$ is upper triangular.*

These three lemmas are sufficient to complete the proof of the theorem. We have $B = XY$, and there exists $T, (n-1) \times (n-1)$, such that $TYXT^{-1}$ is upper triangular. Moreover, the eigenvalues of XY are just 0 and those of YX, or those of $T(YX)T^{-1}$.

Now rewrite $B = (XT^{-1})(TY)$ and set $U = XT^{-1}, V = TY$. That is, $B = UV$ and VU is upper triangular. Letting $\mathbf{u}_i^T$ be the ith column of U and $\mathbf{v}_i$ be the ith row of V gives

$$B = \sum_{i=1}^{n-1} \mathbf{u}_i^T \mathbf{v}_i \quad \text{and} \quad \mathbf{u}_i \cdot \mathbf{v}_j = 0 \quad \text{for} \quad i < j.$$

Since the eigenvalues of B are those of VU, together with 0, these are 0 and $\mathbf{u}_i \cdot \mathbf{v}_i = 1, \dots, n-1$. Hence A is of the requisite form and has the stated property.

This is really only a proof of the "only if" case; however, the "if" case is merely a direct retracement of the above steps.

We give an example of this construction for a 4×4 case:

$$\begin{bmatrix} 1 \\ 0 \\ 1 \\ 0 \end{bmatrix} \begin{bmatrix} 1 & -2 & 1 & 1 \end{bmatrix} + \begin{bmatrix} 2 \\ 1 \\ 2 \\ -1 \end{bmatrix} \begin{bmatrix} -1 & -1 & 1 & 1 \end{bmatrix} + \begin{bmatrix} -3 \\ 2 \\ 1 \\ 2 \end{bmatrix} \begin{bmatrix} 1 & 2 & -1 & 1 \end{bmatrix}$$

$$+1 \cdot \begin{bmatrix} 1 & & & 0 \\ & 1 & & \\ & & 1 & \\ 0 & & & 1 \end{bmatrix} = \begin{bmatrix} -3 & -10 & 6 & 0 \\ 1 & 4 & 3 & 3 \\ 0 & -2 & 3 & 4 \\ 4 & 6 & 0 & 1 \end{bmatrix}$$

has eigenvalues $1, 1+2, 1+(-3), 1+2$ or $1, 3, -2, 3$.

REMARK. The author has used this technique in constructing exercises for students, and found it quite useful for 3×3 and 4×4 cases. However, the restriction $\mathbf{u}_i \cdot \mathbf{v}_j = 0$ for $i < j$ makes it clumsy at higher levels (which are in any case beyond classroom work).

Problem. It would be pleasant to be able also to give a nice formula for the eigenvectors. For the 3×3 case, we obtain

eigenvalue	eigenvector
k	$\mathbf{v}_1 \times \mathbf{v}_2$
$k + \mathbf{u}_1 \cdot \mathbf{v}_1$	$\mathbf{v}_1 \times (\mathbf{u}_1 \times \mathbf{u}_2)\mathbf{u}_1 + (\mathbf{u}_2 \cdot \mathbf{v}_2)\mathbf{u}_2$
$k + \mathbf{u}_2 \cdot \mathbf{v}_2$	$= (\mathbf{u}_2 \cdot \mathbf{v}_2 - \mathbf{u}_1 \cdot \mathbf{v}_1)\mathbf{u}_2 + (\mathbf{u}_2 \cdot \mathbf{v}_1)\mathbf{u}_1$

This is almost neat, but probably will not generalize easily.

References

1. J.W. Archbold, *Algebra*, Pitman, London, 1964.
2. B. Hartley and T.O. Hawkes, *Rings, Modules and Linear Algebra*, Chapman and Hall, London, 1970.
3. M. Newman, *Integral Matrices*, Academic Press, New York, 1972.

J.C. Renaud
University of Papua New Guinea
American Mathematical Monthly **90** (1983), 202–203

Matrices with "Custom Built" Eigenspaces

In a recent note [1], J.C. Renaud poses the question of constructing an $n \times n$ matrix with integer entries and integer eigenvalues. The motivating problem is the need to provide suitable matrices for classroom exercises in linear algebra. He proves an interesting theorem which characterizes such matrices and can be used in their construction.

The motivating problem can be solved pragmatically in a way which may not impress the purist, but will (if thoughtfully applied) provide a supply of suitable classroom exercises and answer the more general motivating question:

> *How can one produce an $n \times n$ matrix possessing each of the following: (a) integer entries; (b) a set of n integer eigenvalues in a specified proportion to each other; (c) a corresponding set of n specified independent eigenvectors having integer entries?*

Such a matrix, A, is diagonalisable and so a nonsingular matrix P exists such that

$$P^{-1}AP = D,$$

where D is a diagonal matrix with entries which are the eigenvalues of A. The matrix P, of course, has the specified eigenvectors as columns. It follows that

$$A = PDP^{-1}.$$

If we begin by constructing matrix P, using the specified n independent eigenvectors as columns, and the matrix D, using a corresponding set of integers in the proportion specified in (b), then the matrix A' produced by the product

$$A' = PDP^{-1}$$

will possess properties (b) and (c) but not, in general, property (a).

This problem can be overcome by a variety of ruses. For example, A' will have integer entries if the determinant of P, $\det P$, is a factor common to each eigenvalue.

A less restrictive approach is to begin with a set of relatively prime eigenvalues and then multiply each entry of the matrix obtained, A', by $\det P$. This will ensure that the resulting matrix, A'', has the desired properties. However, each entry of A'' and each of its eigenvalues may have a factor, f, in common with $\det P$. In that case a more manageable and still suitable matrix A'' can be obtained from A'' by dividing each of its elements by f. If $\{\lambda_i\}_{i=1}^{n}$ is the set of eigenvalues chosen for matrix A', then $\{((\det P)/f)\lambda_i\}_{i=1}^{n}$ will be the set of eigenvalues of matrix A'''. The original set of eigenvectors selected for A' remain the eigenvectors of A'''.

The calculations involved are extensive and require the writing of a suitable computer program. This would invite the input of the order of the matrix, its eigenvalues, and the corresponding eigenvectors. Subroutines for transposing a matrix, inverting a matrix and multiplying matrices would quickly produce the matrix A'. A further subroutine could find $\det P$ and the factor f. These values could then be used to display matrix A''' and its eigenvalues.

Earlier, it was remarked parenthetically, that in using this method to generate suitable classroom exercises, a thoughtful application is needed. Clearly the eigenvectors need to be chosen so that the value of $\det P$ is small, otherwise the magnitude of the elements of A''' may be too large to make pencil and paper calculation convenient. Of interest in this context is an article by Robert Hanson [2], in which an algorithm for constructing matrices which have determinant $+1$ or -1 is established. This can be used in the present situation to construct a convenient matrix P and hence a suitable set of eigenvectors with integer entries. If it is desired, the algorithm could be coded as a further subroutine.

The program will be useful in a variety of teaching and learning situations, including the production of matrices (with integer entries) possessing a set of specified orthogonal eigenvectors. This problem has been investigated by Konrad J. Heuvers [3]. He provides several results which can be used to construct a real $n \times n$ symmetric matrix having prescribed orthogonal eigenvectors and prescribed eigenvalues. The program may also be useful in the search for possible further characterizations of matrices with integer entries and integer eigenvalues.

References

1. J.C.Renaud, Matrices with integer entries and integer eigenvalues, *American Mathematical Monthly* **90** (1983), 202–203.
2. Robert Hanson, Integer matrices whose inverses contain only integers, *Two-Year Coll. Math J.*, **13** (1982), 18–21.
3. Konrad J. Heuvers, Symmetric matrices with prescribed eigenvalues and eigenvectors, *Math. Mag.* **55** (1982), 106–111.

W.P. Galvin
Newcastle, CAE, N.S.W., 2298, Australia
American Mathematical Monthly **91** (1984), 308–309

A Note on Normal Matrices

It is well known that the product of two Hermitian matrices is itself Hermitian if and only if the two matrices commute. The generalization of this statement to normal matrices (A is normal if $A^*A = AA^*$) is not valid, although it is easy to see that if A and B are normal and commute then AB is also normal. A pair of noncommuting unitary matrices will suffice to show that the converse does not hold. In this note we discuss briefly a sufficient condition on a normal matrix A so that for normal B we will have AB normal if and only if A and B commute. It should be observed that if A, B and AB are normal so is BA (see [2]).

A normal matrix A, being unitarily similar to a diagonal matrix, always has the spectral representation

$$(1) \qquad A = \lambda_1 E_1 + \lambda_2 E_2 + \cdots + \lambda_k E_k,$$

where $\lambda_1, \lambda_2, \cdots, \lambda_k$ are the distinct eigenvalues of A and the projectors E_i are uniquely determined polynomials in A such that $E_i, E_j = \delta_{ij} E_i$ and $E_1 + E_2 + \cdots + E_k = I$. The E_i are always Hermitian and B will commute with A if and only if B commutes with each of the E_i (see [1]).

For any complex matrix A we can find a unique positive semi-definite H and a unitary U (not unique if $\det(A) = 0$) such that $A = HU$. This result is a generalization of the polar representation of complex numbers and H is called the Hermitian polar matrix of A. It is not hard to see that H is the unique positive semi-definite square root of AA^*. In [2] Wiegmann proved that if A and B are normal then AB (and hence BA) is normal if and only if each factor commutes with the Hermitian polar matrix of the other.

From the spectral representation of A (1) it follows that

$$(2) \qquad A^* = \overline{\lambda}_1 E_1 + \overline{\lambda}_2 E_2 + \cdots + \overline{\lambda}_k E_k,$$

$$(3) \qquad AA^* = \lambda_1 \overline{\lambda}_1 E_1 + \lambda_2 \overline{\lambda}_2 E_2 + \cdots + \lambda_k \overline{\lambda}_k E_k,$$

and

$$(4) \qquad H = (AA^*)^{1/2} = |\lambda_1| E_1 + |\lambda_2| E_2 + \cdots + |\lambda_k| E_k.$$

If (4) is the spectral representation of H then B commutes with H if and only if B commutes with each E_i, and if B commutes with each E_i then it follows from (1) that B commutes with A. Only if $|\lambda_i| = |\lambda_j|$ for $i \neq j$ could (4) fail to be the spectral representation of H.

We shall say that A has *modularly distinct eigenvalues* if unequal eigenvalues of A have unequal moduli. Our discussion yields the following result.

Theorem. *If A and B are normal and one of the two has modularly distinct eigenvalues, then AB is normal if and only if A and B commute.*

No real matrix with a complex eigenvalue can have this property, but all positive semi-definite and all negative semi-definite Hermitian matrices have modularly distinct eigenvalues; our theorem has the following consequence.

Corollary. If A is positive (negative) semi-definite and B is normal then AB is normal if and only if A and B commute.

References

1. S. Perlis, *Theory of Matrices*, Addison-Wesley, Reading, MA, 1952.
2. N.A. Wiegmann, Normal products of matrices, *Duke Math. J.* **15** (1948), 633–638.

C.G. Cullen
University of Pittsburgh
American Mathematical Monthly **72** (1965), 643–644

Eigenvectors: Fixed Vectors and Fixed Directions (Discovery Exercises)

Except as noted, all matrices, vectors and scalars are real.

1. We say that a square matrix A *fixes* a vector v if $Av = v$.

   a. Show that the zero vector is fixed by every matrix.

   b. Suppose that A fixes a vector v, and that c is a scalar. Does A fix cv?

   c. Suppose that A fixes the vectors v and w, does A fix $v + w$?

   d. Suppose that I_n is the $n \times n$ identity matrix. Does I_n fix any nonzero vectors? Does I_n fix a basis for $\Re^n$?

   e. Suppose that $D = diag(1, 2, 3, \ldots, n)$. Does D fix any nonzero vectors? Does D fix a basis for $\Re^n$? (Hint: Examine D for several small choices of n in formulating your response.)

   f. Suppose that E is the 4×4 matrix $E = diag(1, 1, 1, 0)$. Does E fix any nonzero vectors? Does E fix a basis for $\Re^4$?

   g. What properties must a diagonal matrix have in order to fix a basis for $\Re^n$?

2. We say that a square matrix A *fixes a direction* if there is a nonzero vector v for which Av has the same (or opposite) direction as v. That is, A fixes the direction specified by the nonzero vector v provided there exists a scalar λ such that $Av = \lambda v$.

   a. Explain the relationship between fixing a vector and fixing a direction.

   b. Suppose that A fixes the direction of v, and that c is a scalar. Does A fix the direction of cv? Give a description of what happens geometrically to vectors pointing in a fixed direction when they are multiplied by the matrix A.

   c. Does the matrix I_n defined in Exercise 1d above have any fixed directions that do not correspond to fixed vectors?

   d. Does the matrix D defined in Exercise 1e above have any fixed directions? Does the matrix D have any fixed directions that do not correspond to fixed vectors? What values of λ occur? Find a basis for $\Re^n$ that consists entirely of vectors that correspond to fixed directions.

e. Based on your experience in answering part d, consider this case: If A fixes the direction given by the nonzero vector v, and if A fixes the direction given by the nonzero vector w, must A fix the direction given by the vector $v + w$? If you need to impose additional conditions so that A fixes the direction of $v + w$ when A fixes the directions of v and w, what are the additional conditions?

f. Does the matrix E defined in Exercise 1f have any fixed directions that do not correspond to fixed vectors? What values of λ occur? Find a basis for $\Re^n$ that consists entirely of vectors that correspond to fixed directions.

3. Let $F = \begin{bmatrix} 2 & 1 & 1 & 1 \\ 1 & 2 & 1 & 1 \\ 1 & 1 & 2 & 1 \\ 1 & 1 & 1 & 2 \end{bmatrix}$.

a. Show that $v_1 = [1, -1, 0, 0]^T$, $v_2 = [1, 0, -1, 0]^T$, and $v_3 = [1, 0, 0, -1]^T$ are all fixed vectors for F.

b. Is $v_4 = [1, 1, 1, 1]^T$ a fixed vector for F? Does it correspond to a fixed direction?

c. Let S be the 4×4 matrix whose columns are v_1, v_2, v_3, and v_4. Verify that the columns of S are independent.

d. Relate the columns of FS to the columns of S.

e. Find a matrix M such that $FS = SM$.

f. Compare the matrices $F^k S$ and SM^k for various integers k. Which product requires fewer computations? Does this suggest a reason for finding fixed directions for a matrix? Explain.

4. Let $G = \begin{bmatrix} 0 & 1 \\ 1 & 0 \end{bmatrix}$.

a. Find two independent, fixed directions for G. What scalars λ did you use?

b. Call the direction vectors you found in part a, v_1 and v_2. Let T be the matrix whose columns are your two vectors. Relate the columns of GT to the columns of T.

c. Find a matrix N such that $GT = TN$.

5. Let $H = \begin{bmatrix} 0 & -1 \\ 1 & 0 \end{bmatrix}$.

a. Find a fixed direction for H.

b. If you allow *complex vectors* and *complex scalars*, show that $v_1 = [1, i]^T$ is a fixed direction for H. What is the required complex scalar λ?

c. Find a second fixed direction for H. (Hint: Look for a vector v_2 of the form $v_2 = [1, *]^T$ where $*$ is a complex number other than i.) What is the required complex scalar λ?

d. Show that your vectors v_1 and v_2 are independent, and use them as the columns to form the matrix U. Find a matrix P such that $HU = UP$.

6. Let $K = \begin{bmatrix} 1 & -1 \\ 0 & 1 \end{bmatrix}$.

a. Find a fixed vector v_1 for K.

b. Show that K has no other independent fixed vectors, and that K has no other fixed directions independent from the direction given by v_1.

c. Let V be the matrix whose first column is v_1 and whose second column is $[x, y]^T$. What condition(s) on x and y are necessary in order that the columns of V be independent?

d. Find values for the scalars x and y, where $y \neq 0$, such that $KV = VQ$, where $Q = diag(1, d)$ for some scalar d.

J. Stuart
University of Southern Mississippi
Contributed

On Cauchy's Inequalities for Hermitian Matrices

The purpose of this note is to give a simple and elementary proof for the following theorem, whose proof is usually accomplished by an application of the Courant-Fischer theorem [1,2].

Theorem. *Let A be an n-square Hermitian matrix with eigenvalues*

$$\lambda_1 \geq \cdots \geq \lambda_n,$$

and let B be a k-square principal submatrix of A with eigenvalues $\mu_1 \geq \cdots \geq \mu_k$. Then

$$\lambda_{n-k+s} \leq \mu_s \leq \lambda_s, \qquad s = 1, \ldots, k.$$

Proof. It is sufficient to prove the theorem for $k = n - 1$, i.e., that the eigenvalues of an $(n-1)$-square principal submatrix of A interlace with the eigenvalues of A. The general case follows by applying the result to a chain of matrices $A, B_1, B_2, \ldots, B_{n-k-1}, B$, where B_1 is $(n-1)$-square, B_2 is $(n-2)$-square,..., B_{n-k-1} is $(k+1)$-square and each is a principal submatrix of the preceding one. Consequently, let B be an $(n-1)$-square principal submatrix of A, obtained by the deletion of the qth row and the qth column of A. Let $A = UDU^*$, where $U = (u_{ij})$ is a unitary n-square matrix and $D = \text{diag}(\lambda_1, \ldots, \lambda_n)$. We have

$$[\det(\lambda I - A)]^{-1}\text{adj}(\lambda I - A) = (\lambda I - A)^{-1} = U(\lambda I - D)^{-1}U^*,$$

whence, taking the (q, q) entry in both sides, we obtain

$$\frac{\det(\lambda I - B)}{\det(\lambda I - A)} = \sum_{j=1}^{n} \frac{|u_{qj}|^2}{\lambda - \lambda_j}.$$

The above function is clearly monotonically decreasing at all points of continuity. It follows that it has precisely one zero between two successive poles and such a zero is necessarily an eigenvalue of B. If A has only simple eigenvalues and all $u_{qj} \neq 0$, then $n - 1$ eigenvalues of B interlace strictly with the n eigenvalues of A. It may, of course, happen that A has multiple eigenvalues or that some of the $u_{qj} = 0$. In that case some of the eigenvalues of B will coincide with some of those of A, but the interlacing property is preserved.

References

1. H.L. Hamburger and M.E. Grimshaw, *Linear Transformations in n-Dimensional Vector Space*, Cambridge University, London, 1951.
2. M. Marcus and H. Minc, *Introduction to Linear Algebra*, Macmillan, New York, 1965.

Emeric Deutsch and Harry Hochstadt
Polytechnic Institute of New York
American Mathematical Monthly **85** (1978), 486–487

The Monotonicity Theorem, Cauchy's Interlace Theorem and the Courant-Fischer Theorem

1. Introduction. In this note some important theorems on eigenvalues of Hermitian matrices are reworked from a unified viewpoint of exploiting a simple dimensional identity to obtain easier and quicker proofs. The usual procedure of invoking the minimax characterization or Sylvester's Law of Inertia to prove these results leads to longer proofs (see, for instance, [1, 186–192] or [2, 99–104]).

Our proofs depend on the following simple dimensional identity:

$$\dim(S_1 \cap S_2) = \dim S_1 + \dim S_2 - \dim(S_1 + S_2), \tag{1}$$

where S_1 and S_2 are subspaces of a finite-dimensional vector space. Thus, this note may also be viewed as a collection of good applications of the dimensional identity (1).

Before proceeding, we state the following basic facts used in the subsequent proofs without explicit reference: (a) the eigenvalues of a Hermitian matrix are real and the corresponding eigenvectors may be taken to be orthonormal; (b) letting $\alpha_1 \leq \cdots \leq \alpha_k$ denote a subset of eigenvalues of a Hermitian matrix $\mathcal{A}$ and letting $u_1, \ldots, u_k$ denote an orthonormal set of corresponding eigenvectors, we have $\alpha_1 \leq x^H \mathcal{A} x \leq \alpha_k$ for any x in the span of $u_1, \ldots, u_k$, where $x^H x = 1$. (The symbol "H" denotes conjugate transpose.)

2. The Monotonicity Theorem [1, p. 191].

Let $\mathcal{A}$ and $\mathcal{B}$ be Hermitian and let $\mathcal{A} + \mathcal{B} = \mathcal{C}$. Let the eigenvalues of $\mathcal{A}, \mathcal{B}$, and $\mathcal{C}$ be $\alpha_1 \leq \cdots \leq \alpha_n, \beta_1 \leq \cdots \leq \beta_n$ and $\gamma_1 \leq \cdots \leq \gamma_n$, respectively. Then

$$
\begin{aligned}
&(1) \ \alpha_j + \beta_{i-j+1} \leq \gamma_i, && (i \geq j) \\
&(2) \ \gamma_i \leq \alpha_j + \beta_{i-j+n}, && (i \leq j) \\
&(3) \ \alpha_i + \beta_1 \leq \gamma_i \leq \alpha_i + \beta_n.
\end{aligned}
$$

Proof. Let

$$\mathcal{A}u_i = \alpha_i u_i, \qquad \mathcal{B}v_i = \beta_i v_i, \qquad \mathcal{C}w_i = \gamma_i w_i,$$

$$u_i^H u_j = v_i^H v_j = w_i^H w_j = \delta_{ij}, \quad i, j = 1, \ldots, n.$$

Consider first the case $i \geq j$ and let

$$
\begin{aligned}
S_1 &= \operatorname{span}\{u_j, \ldots, u_n\}, & \dim S_1 &= n - j + 1; \\
S_2 &= \operatorname{span}\{v_{i-j+1}, \ldots, v_n\}, & \dim S_2 &= n - i + j; \\
S_3 &= \operatorname{span}\{w_1, \ldots, w_i\}, & \dim S_3 &= i.
\end{aligned}
$$

Then (1) gives

$$\dim\left(S_1 \cap S_2 \cap S_3\right) \geq \dim S_1 + \dim S_2 + \dim S_3 - 2n = 1.$$

This assures the existence of an $x \in S_1 \cap S_2 \cap S_3$ such that $x^H x = 1$. For this x we have

$$\alpha_j + \beta_{i-j+1} \leq x^H \mathcal{A} x + x^H \mathcal{B} x = x^H \mathcal{C} x \leq \gamma_i,$$

proving (1). Application of (1) to $(-\mathcal{A}) + (-\mathcal{B}) = -\mathcal{C}$ proves (2). Setting $i = j$ in (1) and (2) gives (3).

3. Cauchy's Interlace Theorem [1, p. 186].

Let

$$\mathcal{A} = \begin{bmatrix} \mathcal{B} & \mathcal{C} \\ \mathcal{C}^H & \mathcal{D} \end{bmatrix}$$

be an $n \times n$ Hermitian matrix, where $\mathcal{B}$ has size $m \times m$ ($m < n$). Let eigenvalues of $\mathcal{A}$ and $\mathcal{B}$ be $\alpha_1 \leq \cdots \leq \alpha_n$ and $\beta_1 \leq \cdots \leq \beta_m$, respectively. Then

$$\alpha_k \leq \beta_k \leq \alpha_{k+n-m}, \qquad k = 1, \dots, m.$$

Proof. Let

$$\begin{aligned}
\mathcal{A} u_i &= \alpha_i u_i, \quad u_i^H u_j = \delta_{ij}, \quad i, j = 1, \dots, n, \\
\mathcal{B} v_i &= \beta_i v_i, \quad v_i^H v_j = \delta_{ij}, \quad i, j = 1, \dots, m, \\
w_i &= \begin{bmatrix} v_i \\ 0 \end{bmatrix}, \qquad\qquad\quad i = 1, \dots, m.
\end{aligned}$$

Let $1 \leq k \leq m$ and let

$$\begin{aligned}
S_1 &= \operatorname{span}\{u_k, \dots, u_n\}, \quad \dim S_1 = n - k + 1; \\
S_2 &= \operatorname{span}\{w_1, \dots, w_k\}, \quad \dim S_2 = k.
\end{aligned}$$

Again by §1(1), the existence of an $x \in S_1 \cap S_2$ such that $x^H x = 1$ is assured and we have

$$\alpha_k \leq x^H \mathcal{A} x \leq \beta_k.$$

Application of this inequality to $-\mathcal{A}$ gives $\beta_k \leq \alpha_{k+n-m}$.

4. The Courant-Fischer Theorem (Minimax Characterization) [1, p. 188].

Let $\mathcal{A}$ be Hermitian and let $\alpha_1 \leq \cdots \leq \alpha_n$ be the eigenvalues of $\mathcal{A}$. Then for $k = 1, \dots, n$,

$$\begin{aligned}
\alpha_k &= \min_{S^k} \max \left\{ v^H \mathcal{A} v : v \in S^k, v^H v = 1 \right\} \\
&= \max_{S^{k-1}} \min \left\{ v^H \mathcal{A} v : v \perp S^{k-1}, v^H v = 1 \right\},
\end{aligned}$$

where S^k denotes an arbitrary k-dimensional subspace of complex n-vectors.

Proof. Let

$$\mathcal{A}u_i = \alpha_i u_i, \quad u_i^H u_j = \delta_{ij}, \qquad j = 1, \ldots, n.$$

Let

$$S_1 = \text{span}\{u_k, \ldots, u_n\} \quad \text{and} \quad S_2 = S^k, \text{ (any } k\text{-dimensional subspace)}.$$

Then §1(1) guarantees the existence of an $x \in S_1 \cap S^k, x^H x = 1$, giving $x^H \mathcal{A}x \geq \alpha_k$.

On the other hand, for any $u \in \text{span}\{u_1, \ldots, u_k\}$, a k-dimensional subspace, we have $u^H \mathcal{A}u \leq \alpha_k$ and $u_k^H \mathcal{A}u_k = \alpha_k$, proving the first equality of the theorem.

To prove the second, choose

$$S_1 = \text{span}\{u_1, \ldots, u_k\}, \qquad S_2 = (S^{k-1})^{\perp},$$

and proceed in a similar line of argument as above.

References

1. B.N. Parlett, *The Symmetric Eigenvalue Problem*, Prentice-Hall, 1980.
2. J.H. Wilkinson, *The Algebraic Eigenvalue Problem*, Oxford, 1965.

Yasuhiko Ikebe, Toshiyuki Inagaki, and Sadaaki Miyamoto
University of Tsukuba, Japan
American Mathematical Monthly **94** (1987), 352–354

The Power Method for Finding Eigenvalues on a Microcomputer

Introduction. Computers are becoming more available for classroom use in mathematics and quality mathematics software is beginning to appear. It is an appropriate time to look into the ways that these tools can be used to explore mathematics. Microcomputers can be used for much more than merely drill and practice. They can be used to introduce a flavor of discovery into courses. Students can discover results and view theorems in a pedagogical manner that has not been possible until this era. With all the computational power needed at their fingertips (power that was only available to research workers in the past), students can examine many situations quickly and focus on the behavior of methods or models. The computer introduces an element of surprise. Things do not always work the way they are expected to. This paper discusses such a situation in the application of the power method to compute the dominant eigenvalue and a corresponding eigenvector of a matrix. It illustrates how much is often learned when things "go wrong."

The Power Method [1]. Numerical techniques exist for evaluating certain eigenvalues and eigenvectors of various types of matrices. The power method is a straightforward iterative method that leads to the dominant eigenvalue (if it exists) and a corresponding eigenvector. It is often taught in linear algebra and numerical methods courses.

The dominant eigenvalue is the one with the largest absolute value. We remind the reader of the power method at this time.

Let A be an $n \times n$ matrix having n linearly independent eigenvectors and a dominant eigenvalue λ. Let X_0 be an arbitrarily chosen initial column vector having n components. If X_0 has a nonzero component in the direction of an eigenvector for λ, then the sequence

$$X_1 = AX_0, \quad X_2 = A\hat{X}_1, \quad X_3 = A\hat{X}_2, \ldots, \quad X_k = A\hat{X}_{k-1}, \ldots$$

will approach an eigenvector for λ. Here $\hat{X}_k$ is a normalized form of X_k, obtained by dividing each component of X_k by the absolute value of its largest component.

Furthermore, the sequence

$$\frac{\hat{X} \cdot A\hat{X}_1}{\hat{X}_1 \cdot \hat{X}_1}, \ldots, \frac{\hat{X}_k \cdot A\hat{X}_k}{\hat{X}_k \cdot \hat{X}_k}, \ldots$$

will approach the dominant eigenvalue.

The power method is not the most efficient numerical method for computing eigenvalues and eigenvectors; convergence can be extremely slow. However, it is the easiest to prove and the most commonly taught in introductory courses.

81

Construction of a Test Case. Let us construct a 3×3 matrix that has known eigenvalues and eigenvectors for testing the power method. A similarity transformation CAC^{-1} performed on a diagonal matrix A will lead to a matrix B having the diagonal elements of A as eigenvalues and having the columns of C as eigenvectors [2].

Let

$$B = CAC^{-1} = \begin{pmatrix} 1 & 0 & 1 \\ 0 & 1 & 1 \\ 1 & 1 & 1 \end{pmatrix} \begin{pmatrix} 1 & 0 & 0 \\ 0 & 4 & 0 \\ 0 & 0 & 6 \end{pmatrix} \begin{pmatrix} 1 & 0 & 1 \\ 0 & 1 & 1 \\ 1 & 1 & 1 \end{pmatrix}^{-1} = \begin{pmatrix} 6 & 5 & -5 \\ 2 & 6 & -2 \\ 2 & 5 & -1 \end{pmatrix}.$$

The eigenvalues of B are thus known to be $1, 4, 6$ with corresponding eigenvectors

$$\begin{pmatrix} 1 \\ 0 \\ 1 \end{pmatrix}, \begin{pmatrix} 0 \\ 1 \\ 1 \end{pmatrix}, \begin{pmatrix} 1 \\ 1 \\ 1 \end{pmatrix}.$$

For convenience we shall henceforth write the eigenvectors as row vectors.

All computation, such as the similarity transformation above, is carried out by students in a linear algebra class using menu driven software [3]. The students do no programming. The similarity transformation is carried out using a matrix inverse program to compute C^{-1} and then using a matrix multiplication program twice to compute AC^{-1} and then $C(AC^{-1})$. All intermediate results in the computation of CAC^{-1} are saved on the computer for use in the next stage.

Let us verify that B does indeed have the above eigenvalues and corresponding eigenvectors. This can be conveniently done using the following multiplication:

$$\begin{pmatrix} 6 & 5 & -5 \\ 2 & 6 & -2 \\ 2 & 5 & -1 \end{pmatrix} \begin{pmatrix} 1 & 0 & 1 \\ 0 & 1 & 1 \\ 1 & 1 & 1 \end{pmatrix} = \begin{pmatrix} 1 & 0 & 6 \\ 0 & 4 & 6 \\ 1 & 4 & 6 \end{pmatrix}.$$

$$\underset{B}{\uparrow} \qquad \underset{\text{eigenvectors}}{\nwarrow \uparrow \nearrow} \qquad \underset{\substack{\text{1, 4 and 6 times} \\ \text{the eigenvectors}}}{\nwarrow \uparrow \nearrow}$$

Thus $(1, 0, 1), (0, 1, 1)$ and $(1, 1, 1)$ are eigenvectors of B corresponding to the eigenvalues $1, 4$ and 6. B can now be used to test the power method. The method should result in the dominant eigenvalue 6 and the corresponding eigenvector $(1, 1, 1)$.

Applying the Power Method. Let us use the vector $X = (1, 2, 3)$ as the initial vector for the power method. This is a popular initial vector with students for iterative methods! The method, using 15 iterations, gives 3.99926723 as the dominant eigenvalue with corresponding eigenvector

$$(2.94530762 \times 10^{-8}, .999999996, .999999996).$$

These are approximations for the second eigenvalue 4 and corresponding eigenvector $(0, 1, 1)$. This is an opportunity for students to examine the conditions of the power method—to find out why the expected results have not occurred.

There are three assumptions made in the power method. The first is that there are three (in this case) linearly independent eigenvectors, the second that there exists a dominant eigenvalue, the third that the initial vector has a nonzero component in the direction of the dominant eigenvector. The first two conditions are satisfied. B has three linearly independent eigenvectors and a dominant eigenvalue, namely 6. Thus the third condition involving a nonzero component in the direction of a dominant eigenvector must be violated. Let us write $(1, 2, 3)$ as a linear combination of eigenvectors. We get

$$(1, 2, 3) = 1(1, 0, 1) + 2(0, 1, 1) + 0(1, 1, 1).$$

The initial vector $(1, 2, 3)$ does indeed have a zero component in the direction of the dominant eigenvector $(1, 1, 1)$. The conditions of the method do not hold. We cannot expect convergence to the dominant eigenvalue.

The next step, of course, is to investigate the convergence to the second eigenvalue. Is this to be expected in general if the initial vector has zero component in the direction of the dominant eigenvector? The proof of the power method is straightforward and students can easily see why, on selecting an initial vector having zero component in the direction of the dominant eigenvector, convergence will take place to the second eigenvalue and corresponding eigenvector if the initial vector has a nonzero component in this direction. This observation suggests how the power method can be used to determine further eigenvalues and eigenvectors once the dominant eigenvalue and eigenvector have been found.

Finally, however, if the power method is continued beyond the 15th iteration, divergence from the eigenvalue 4 takes place with gradual convergence towards the dominant eigenvalue 6. For example after 60 iterations, the method gives 5.54002294 while after 100 iterations it gives 5.99999995. The theory of the power method does not predict such a phenomenon. It predicts convergence to the eigenvalue 4. This is an opportunity to discuss the effect of round-off errors that occur when such methods are executed on computers. At the 16th iteration, X_{16} has a significant nonzero component in the direction of the dominant eigenvector $(1, 1, 1)$ due to round-off errors on the computer. We are back with all the original conditions of the power method being satisfied. X_{16} is the new initial vector and convergence takes place to the dominant eigenvalue and eigenvector beyond this point. The convergence is extremely slow due to the small component of X_{16} in the direction of the dominant eigenvector.

The geometrical interpretation of the above is of interest. The initial vector $(1, 2, 3)$ lies in the subspace spanned by the eigenvectors $(1, 0, 1)$ and $(0, 1, 1)$. The theory of the power method shows that convergence should take place within this subspace to the dominant eigenvalue and eigenvector of this subspace, namely 4 and $(0, 1, 1)$. In our case however, round-off errors cause vectors to stray out of this two-dimensional subspace. This causes vectors beyond the 16th to be gradually dragged toward the eigenvector $(1, 1, 1)$ which lies outside the subspace.

Comments. This example arose by chance in a linear algebra class. An Apple microcomputer was used to carry out the computations using special linear algebra software. The matrix C was introduced as one having a clean inverse. The matrix A and initial guess were supplied by students. The class (and the instructor!) struggled together to interpret the output. The matrix B is now given to students as a regular assignment on the power method. They are asked to interpret the output. It is this type of experimentation that the computer introduces into the teaching of mathematics.

Let us summarize the benefits that have been achieved through this example.

1. The student has applied the power method in the environment that it is meant to be applied, namely on the computer.
2. The importance of conditions under which the method holds has been stressed. The conditions are part of the method—a part that is often overlooked.
3. The student has had the experience of modifying the method to arrive at a generalization.
4. The effect of round-off errors on computers has been emphasized.
5. The importance of interpreting and checking computer output has been illustrated. The student could have assumed that 4 was the dominant eigenvalue if the method had been applied to 15 iterations.
6. The importance of using test cases, with known results, to understand the behavior of algorithms before entering upon the unknown has been vividly illustrated.

The time is now ripe to start collecting examples such as the above for use in the undergraduate classes. These examples can be used to supplement standard course material, adding an element of mathematical exploration. Students now have the opportunity to see dimensions of the field that could not be revealed to previous generations. At the same time it is important to stress that the computer should be used sparingly and naturally. This is a new teaching medium. We are all going to have to discover when and how to use it most effectively.

The authors are interested in starting a collection of such "mathematical experiments" with the idea of making this collection available to the mathematical community. Readers who have discovered such examples are encouraged to contact us.

References

1. Gareth Williams, *Linear Algebra with Applications*, Allyn and Bacon, Boston, MA, 1984, p. 411.
2. Ibid., p. 361.
3. Gareth Williams and Donna J. Williams, Allyn and Bacon, Linear Algebra Computer Companion, Boston, MA, 1984. (Software package consisting of two diskettes and a manual.)

Gareth Williams   Donna Williams
Stetson University   Rollins College
American Mathematical Monthly **93** (1986), 562–564

Singular Values and the Spectral Theorem

The action of an arbitrary real or complex $m \times n$ matrix is not all arbitrary: it takes a suitable orthonormal basis of n-space into some orthogonal set in m-space; consequently it maps the unit n-ball onto a (possibly lower-dimensional) ellipsoid. This is the geometric view of the Singular Value Theorem, which is not only one of the nicest matrix theorems to state and to visualize but also one of the easiest to prove and to apply. In any introductory course on matrices it deserves a place near the center. Theorem 1 below is a pure matrix version of this result. One of its main consequences, Theorem 2 below, leads straight to the orthogonal diagonalization of certain matrices, i.e. the Spectral Theorem.

It is customary to prove Theorem 2 independently, applying either the Fundamental Theorem of Algebra to a characteristic polynomial or Lagrange Multipliers to a quadratic form, and even to deduce Theorem 1 from *it*. The analytic equipment needed for the following proof is more modest: it suffices to know that a continuous real-valued function on any compact set has a maximum. For the sake of simplicity the argument will first be given for *real* matrices and then (trivially) extended to complex ones. To prevent any suspicion that extraneous subtleties are tacitly used, the prefix "eigen" will be avoided.

Theorem 1. *Let $A \neq 0$ be an $m \times n$ matrix. Then there exist orthogonal matrices M and N such that*

$$MAN = \begin{bmatrix} D & 0 \\ 0 & 0 \end{bmatrix} \quad where \quad D = \begin{bmatrix} d_1 & & 0 \\ & \ddots & \\ 0 & & d_r \end{bmatrix} \quad with \quad d_i \geq d_{i+1} > 0.$$

Proof. Let $O(k)$ be the set of all $k \times k$ orthogonal matrices. For $M \in O(m)$ and $N \in O(n)$, let $\alpha(M, N)$ stand for the entry in the first row and first column of MAN. Clearly, α is a continuous function of the pair (M, N) and, therefore, attains a maximal value $d_1 > 0$ on the set $O(m) \times O(n)$, which is closed and bounded in a suitable Euclidean space. Say $d_1 = \alpha(M_1, N_1)$. Then

$$M_1 A N_1 = \begin{bmatrix} d_1 & X \\ Y & A' \end{bmatrix},$$

where A' is an $(m - 1) \times (n - 1)$ matrix. It turns out that X and Y are actually *zero* rows and columns, respectively. Indeed, if X were nontrivial, the first row ρ_1 of $M_1 A N_1$ would have length $d > d_1$. Then one could multiply on the right by the reflection H which takes ρ_1 to $[d, 0, \ldots, 0]$ and create a value $\alpha(M, N) = d > d_1$. For similar reasons, involving columns and left multiplication, it follows that $Y = 0$. Finally, none of the entries of A' can exceed

d_1 in absolute value (because each of them could be permuted to the upper left), and this is true for all the possible forms of A'. An obvious induction now finishes the proof.

The positive numbers $d_1 \geq \cdots \geq d_r$, are known as the *singular values* of A.

Remarks. The $n \times n$ matrix $A^{\#} = A^T A$ has a very simple effect on the columns $v_1, \ldots, v_n$ of N; in fact $A^{\#} v_i = \mu_i v_i$, where $\mu_i = d_i^2$ for $i \leq r$, and $= 0$ beyond. In matrix terms, this says that $A^{\#} N = N \Delta$, where Δ is a diagonal matrix with diagonal entries μ_i as described. It is proved by putting $\Delta = (MAN)^T (MAN)$, which is just $N^T A^{\#} N$.

To prove *uniqueness* of the singular values d_i it clearly suffices to characterize the μ_i as being the only numbers such that $(A^{\#} - \mu I)v = 0$ for some $v \neq 0$. Indeed, setting $v = \sum a_i v_i$, one gets $(A^{\#} - \mu I)v = \sum a_i(\mu_i - \mu)v_i$, which is never 0, unless μ is one of the μ_i. More geometrically, the d_i can also be retrieved from the image under A of the appropriate unit sphere.

If $m = n$, the theorem is often written as $A = SR$, with S positive semidefinite and $R = M^T N^T$ orthogonal. It is then called the polar decomposition.

Theorem 2. *Every real symmetric $n \times n$ matrix A has an invariant one-dimensional subspace.*

Proof. Let $v \neq 0$ be one of the columns of N, so that $(A^{\#} - \mu I)v = 0$, as in the remarks above. The symmetric nature of A makes $A^{\#} = A^T A = A^2$ and

$$0 = (A^{\#} - \mu I)v = (A + \lambda I)(A - \lambda I)v,$$

where $\lambda^2 = \mu$. If $(A - \lambda I)v = w \neq 0$, then w generates such a line; otherwise v does.

Remarks. For symmetric A it is trivial to show that the orthocomplement of any invariant subspace is itself invariant. Hence, by induction, Theorem 2 provides a set of n mutually orthogonal invariant lines, thus proving the spectral theorem for symmetric matrices. Moreover, if B is symmetric and commutes with A, it defines a symmetric operator on the non-zero kernel of $A - \lambda I$. Hence, the two matrices have a *common* invariant line, and again, by induction, a complete orthogonal set of such lines.

If the use of linear transformations is didactically impracticable, the spectral theorem can be stated in terms of an orthogonal U making $U^T AU$ diagonal. Then the induction will hinge on partial diagonalizations $V^T AV$, where the columns of V form an orthonormal basis of an invariant subspace and its orthocomplement.

Everything said after the first sentence of Theorem 1 and before the present one applies *verbatim* to complex matrices, if one changes "orthogonal" to "unitary," "symmetric" to "hermitian," and replaces the transpose A^T by its complex conjugate A^* (adjoint of A). Writing every complex matrix as $C = A + iB$, with A and B hermitian, one can clearly extend the spectral theorem to all C for which A and B commute, i.e., where C commutes with its adjoint $A - iB$. These are, of course, the *normal* matrices.

K. Hoechsmann
University of British Columbia, Vancouver
American Mathematical Monthly **97** (1990), 413–414

The Characteristic Polynomial of a Singular Matrix

Let $A = [a_{ij}]$ be an $n \times n$ matrix of rank $r < n$. Then 0 is a characteristic value of A of multiplicity at least $n - r$. Thus, if $\phi(\lambda) = |\lambda I - A|$ is the characteristic polynomial of A, $\phi(\lambda) = \lambda^{n-r}\psi(\lambda)$ where $\psi(\lambda)$ is a polynomial of degree r. We ask the question: *Is there any way to find $\psi(\lambda)$ without first finding $\phi(\lambda)$?* The purpose of this note is to give an affirmative answer and to show that $\psi(\lambda)$ is the characteristic polynomial of an $r \times r$ matrix D which is related to A in a simple way.

Since A is of rank $r < n$, it may be represented (actually in many ways) as a product BC of two $n \times n$ matrices where the last $n - r$ columns of B and the last $n - r$ rows of C consist of zeros. For example, the first r rows of C could be any r independent rows of A; B would then contain the proper multipliers to generate A.

Now CB will be of the form diag $\{D, 0\}$ where D is an $r \times r$ matrix. By a theorem originally stated by Sylvester [2] and proved in many modern textbooks (e.g. [1,p. 23]), BC and CB have the same characteristic polynomial. Thus

$$\phi(\lambda) = |\lambda I_{n \times n} - A| = |\lambda I_{n \times n} - CB| = \lambda^{n-r}|\lambda I_{r \times r} - D| = \lambda^{n-r}\psi(\lambda).$$

If D is of rank less than r, the procedure could be repeated.

The result would seem to be particularly useful when r is small in comparison to n. We illustrate for $r = 1$ and $r = 2$, noting that the representation $A = BC$ above is equivalent to $a_{ij} = \sum_{k=1}^{r} b_{ik}c_{kj}, 1 \leq i, j \leq n$, and that the i,jth element of D is then given by $d_{ij} = \sum_{k=1}^{n} c_{ik}b_{kj}, 1 \leq i, j \leq r$.

Example 1: $(r = 1)$. If $a_{ij} = b_i c_j$, then

$$|\lambda I - A| = \lambda^{n-1}\left(\lambda - \sum_{i=1}^{n} b_i c_i\right).$$

Example 2: $(r = 2)$. If $a_{ij} = b_i c_j + d_i e_j$, then

$$|\lambda I - A| = \lambda^{n-2}\begin{vmatrix} \lambda - \sum_{i=1}^{n} b_i c_i & -\sum_{i=1}^{n} c_i d_i \\ -\sum_{i=1}^{n} b_i e_i & \lambda - \sum_{i=1}^{n} d_i e_i \end{vmatrix}.$$

Acknowledgements. Improvements in exposition resulted from the referee's comments. This work was supported by the U.S. Air Force Office of Scientific Research under Grant No. AF-AFOSR 35–65.

References

1. C.C. MacDuffee, *The theory of matrices*, Chelsea, New York, 1956.
2. J.J. Sylvester, *Philos. Mag.* **16** (1883), 267–269.

D.E. Varberg
Hamline University
American Mathematical Monthly **75** (1968), 506

Characteristic Roots of Rank 1 Matrices

The purpose of this note is to prove a result which generalizes popular elementary Problem E 1859 (this MONTHLY, 74 (1967) 598, 724).

Theorem. *Let A denote an $n \times n$ rank 1 matrix over any field. When Trace (A) is nonzero, it is a simple characteristic root of A and 0 is a root of multiplicity $n-1$. Otherwise, the only characteristic root of A is 0.*

Proof. When $n = 1$ the result is trivial. We first show directly that Trace (A) is a characteristic root of A. Suppose the row vectors of A are $A_1, \cdots, A_n$. Since A is not zero, A has a nonzero row, say A_k. Then there are scalars $\lambda_1, \cdots, \lambda_n$ with $\lambda_k = 1$ for which

$$A_i = \lambda_i A_k \qquad (i = 1, \cdots, n).$$

Let $A_k = (a_1, \cdots, a_n)$. Then A may be written as $[\lambda_i a_j]$. For any n-vector $x = (x_1, \cdots, x_n), (A_k, x)$ will denote the usual inner product $\sum_{i=1}^{n} a_i x_i$. In particular, $(A_k, \omega) = \mathrm{Trace}(A)$, where ω is the nonzero vector $(\lambda_1, \cdots, \lambda_n)$. A simple calculation shows

$$Ax^T = (A_k, x)\omega^T.$$

Thus, ω is a characteristic vector of A corresponding to $(A_k, \omega) = \mathrm{Trace}(A)$ and $Ax^T = 0$ only if $(A_k, x) = 0$.

To determine the multiplicities of 0 and Trace (A), we note that A must be similar to a matrix all of whose rows except the kth are zero. For, if A is not already in this form, $\lambda_j \neq 0$ for some $j \neq k$ and $C = B_{jk}(-\lambda_j)AB_{jk}(\lambda_j)$ is a matrix similar to A with jth row zero. $B_{ij}(\lambda)$ here denotes the elementary matrix $I + \lambda E_{jk}$, where E_{ij} is the matrix with entry (i, j) equal to 1 and zeros elsewhere. Since C is also a rank 1 matrix we may repeat the argument as necessary with a possibly different set of scalars $\lambda_1, \cdots, \lambda_n$ to obtain a matrix $C = [c_{ij}]$ which is similar to A with only $C_k \neq 0$. Since the characteristic equation of C is $\lambda^{n-1}(\lambda - c_{kk}) = 0$, 0 is a root of A with multiplicity $\geq n - 1$. Since one root of the characteristic equation is Trace (A), the multiplicity of zero is $n - 1$ if Trace (A) is nonzero and otherwise is n. This completes the proof.

The fact that Trace (A) is a characteristic root is also seen by noting that the trace is a similarity invariant and Trace $(C) = c_{kk}$ is a root of C with characteristic vector $e_k = (\delta_{1k}, \cdots, \delta_{nk})$.

Larry Cummings
University of Waterloo
American Mathematical Monthly **75** (1968), 1105

A Method for Finding the Eigenvectors of an $n \times n$ Matrix Corresponding to Eigenvalues of Multiplicity One

Many texts [1],[2] and introductory courses on linear algebra introduce the adjoint matrix (the transpose of the cofactor matrix) to determine an expression for the inverse of an $n \times n$ invertible matrix. The purpose of this report is to show, using elementary linear algebra techniques, that the adjoint matrix can also be used to find the eigenvectors of an $n \times n$ matrix corresponding to eigenvalues of multiplicity one. An advanced analysis of our results lying outside the scope of an introductory linear algebra course can be found in the text *Matrix Theory* by F.R. Gantmacher [3].

As mentioned, the introductory linear algebra student's first encounter with the adjoint matrix appears in the expression [1],[2]

$$(1) \qquad A[adj(A)] = (\det A)I,$$

where $adj(A)$ is the adjoint matrix of an $n \times n$ matrix A, and $\det A$ is its determinant. When $\det A \neq 0$, Equation (1) leads to the usual formula,

$$A^{-1} = [\det A]^{-1}[adj(A)]$$

for the inverse of the matrix. Equation (1), however, is usually not explored in an introductory course for the case when $\det A = 0$. We will consider one implication of this case. When $\det A = 0$, Equation (1) becomes

$$(2) \qquad A[adj(A)] = \mathbf{0}.$$

If λ_i is an eigenvalue of a matrix H (making $\det(\lambda_i I - H) = 0$), then the substitution $A = \lambda_i I - H$ into Equation (2) gives $(\lambda_i I - H)[adj(\lambda_i I - H)] = \mathbf{0}$ or

$$(3) \qquad H[adj(\lambda_i I - H)] = \lambda_i [adj(\lambda_i I - H)].$$

The pth column of this matrix equation is

$$(4) \qquad H[adj(\lambda_i I - H)]_p = \lambda_i [adj(\lambda_i I - H)]_p,$$

which shows that any nonzero column $[adj(\lambda_i I - H)]_p$ of $adj(\lambda_i I - H)$ is an eigenvector for H corresponding to the eigenvalue λ_i.

As an example of this method consider the 3×3 matrix

$$H = \begin{bmatrix} 0 & 1 & 0 \\ 0 & 0 & 1 \\ 4 & -17 & 8 \end{bmatrix}$$

which has $\lambda_i = 4$ as an eigenvalue of multiplicity one. Then

$$4I - H = \begin{bmatrix} 4 & -1 & 0 \\ 0 & 4 & -1 \\ -4 & 17 & -4 \end{bmatrix} \quad \text{and} \quad adj(4I - H) = \begin{bmatrix} 1 & -4 & 1 \\ 4 & -16 & 4 \\ 16 & -64 & 16 \end{bmatrix}.$$

Since $[adj(4I - H)]_{11}$ is not zero, we can take $p = 1$, giving $\begin{bmatrix} 1 & 4 & 16 \end{bmatrix}^T$ as an eigenvector for H corresponding to the eigenvalue $\lambda_i = 4$. It is usually not necessary to calculate the entire $adj(\lambda_i I - H)$ matrix. In the example above, once we find that $[adj(4I - H)]_{11}$ is not zero, then we need only find the 1st column of $adj(\lambda_i I - H)$ when obtaining the eigenvector. In fact, for a 2×2 or 3×3 matrix this technique is much faster and easier to program on a computer than the usual methods of computing eigenvectors.

The only complication to the method above arises when $adj(\lambda_i I - H)$ is the zero matrix. To address this complication, we use the following result from linear algebra for the derivative of $\det(\lambda I - H)$[**4**]:

$$(5) \qquad (d/d\lambda)[\det(\lambda I - H)] = \text{Trace}\left\{ adj(\lambda I - H) \right\}.$$

We shall apply this result to the eigenvalue polynomial

$$\det(\lambda I - H) = \prod_{j=1}^{n} (\lambda - \lambda_j),$$

where $\lambda_1, \lambda_2, \lambda_3, \ldots, \lambda_n$ are the eigenvalues of H. Taking the derivative of both sides of this equation with respect to λ and using Equation (5) leads to

$$(6) \qquad \text{Trace}\left\{ adj(\lambda I - H) \right\} = \sum_{k=1}^{n} \prod_{\substack{j=1 \\ j \neq k}}^{n} (\lambda - \lambda_j).$$

By setting $\lambda = \lambda_i$ in Equation (6) we obtain

$$(7) \qquad \text{Trace}\left\{ adj(\lambda_i I - H) \right\} = \sum_{j=1}^{n} [adj(\lambda_i I - H)]_{jj} = \prod_{\substack{j=1 \\ j \neq i}}^{n} (\lambda_i - \lambda_j).$$

Therefore if λ_i is an eigenvalue of multiplicity one, the right-hand side of Equation (7) is nonzero. It follows that there exists a value of p between and 1 and n for which $[adj(\lambda_i I - H)]_{pp}$ is nonzero, and hence $adj(\lambda_i I - H) \neq \mathbf{0}$.

To summarize: Let H be an $n \times n$ matrix with eigenvalues $\lambda_1, \lambda_2, \lambda_3, \ldots, \lambda_n$. Further, let λ_i for some i have multiplicity one. Then there exists a value of p such that the pth column of $adj(\lambda_i I - H)$ is an eigenvector for H corresponding to λ_i. Moreover, a value of p to use is one for which $[adj(\lambda_i I - H)]_{pp}$ is not zero (i.e., we need only consider the cofactors of the diagonal elements of $\lambda_i I - H$ when determining p). Of course, once one eigenvector corresponding to an eigenvalue of multiplicity one is known, all others are just multiples of it.

For the case when the multiplicity of λ_i is greater than one, it can be shown [**3**] that the matrix $adj(\lambda_i I - H)$ is nonzero provided the eigenspace of eigenvectors corresponding to the eigenvalue λ_i had dimension exactly equal to one. However, in this case $adj[(\lambda_i I - H)]_{pp}$ may be zero for all p, and so the choice of a suitable value of p may be more difficult.

Acknowledgement. I wish to thank Dr. C. Rorres for reviewing this manuscript.

References

1. Howard Anton, *Elementary Linear Algebra*, 4th ed., Wiley, New York, 1984.
2. Bernard Kolman, *Introductory Linear Algebra with Applications*, 3rd ed., Macmillan, New York, 1984.
3. F.R. Gantmacher, *Matrix Theory*, Vol. I, Chelsea, New York, 1960.
4. Charles G. Cullen, *Matrices and Linear Transformations* (Theorem 8–15, page 183), Addison-Wesley, Reading, MA, 1966.

M. Carchidi
Drexel University
American Mathematical Monthly **93** (1986), 647–649

PART 4
Geometry

Introduction

The interactions between linear algebra and geometry are many and varied. Much of the elementary linear algebra can be described as "using the ideas of geometry to see what to do, and using the formulas of algebra to actually do it precisely." Gram-Schmidt projections, and the Cauchy-Schwarz inequality, and Standard Least Squares Problems (the titles of three of the articles in this section.) are beautiful and simple examples of how natural geometric notions in R^2 and R^3 carry over to R^n with algebraic techniques that work for $n = 2, 3, \ldots$. Orthogonality is meaningful and useful in spaces for which no protractor exists.

Gram-Schmidt Projections

Given a linearly independent set in an inner product space, the Gram-Schmidt orthonormalization process is an algorithm that replaces it with an orthonormal set that spans the same subspace. Gram-Schmidt is mentioned in many linear algebra books, e.g., [4, 0.6.4]. If $\{x_1, \ldots, x_n\}$ is a linearly independent set in $\Re^m, m \geq n$, the sequential calculation of the resulting orthonormal set $\{z_1, \ldots, z_n\}$ is given by

$$y_k = x_k - \sum_{i=1}^{k-1} \left(z_i^T x_k \right) z_i,$$

$$z_k = \frac{y_k}{\left(y_k^T y_k \right)^{\frac{1}{2}}}, \quad \text{for} \ \ k = 1, \ldots, n. \tag{1}$$

If X is the m-by-n matrix with columns $x_1, \ldots, x_n$ and Z the m-by-n matrix with columns $z_1, \ldots, z_n$, this results in the factorization $X = ZR$, with Z orthogonal and R upper triangular. We work with $x_i \in \Re^m$, but note that with obvious modifications, our results hold for $x_i \in \mathbb{C}^m$; for example, the matrix Z is then unitary.

The factorization of an m-by-n matrix A into a product QR, where Q is an m-by-n matrix with orthonormal columns and R is upper triangular, is called a QR factorization of A; see [4, 2.6]. This factorization is used, for example, in numerical methods for solving least squares problems and computing eigenvalues and singular values; see [2]. When A has full column rank, it has a unique factorization $A = QR$, where Q is an m-by-n matrix with orthonormal columns, and R is an n-by-n upper triangular matrix with positive main diagonal entries.

Here we consider a fixed m-by-n real matrix A having full column rank, and define a sequence of m-by-m matrices $B^{(k)}, k = 0, \ldots, n-1$, which we note are the linear transformations that implement the Gram-Schmidt process. This provides an alternative (and we believe novel, though elementary) way of viewing the process and of building up the matrix Q of the QR factorization. This is an outgrowth of our detailed study of the combinatorial structure of the matrices Q and R of this factorization [3].

Let a_i, q_i denote the ith column of matrix A, Q, respectively, and let I_m denote the m-by-m identity matrix. Given an m-by-n matrix A, we claim that the Gram-Schmidt process applied to the column vectors of A can be written in terms of matrix transformations as follows:

$$y_k = B^{(k-1)} a_k,$$
$$q_k = \frac{y_k}{\left(y_k^T y_k \right)^{\frac{1}{2}}}, \quad \text{for} \ \ k = 1, \ldots, n, \tag{2}$$

where

$$B^{(0)} = I_m, \quad B^{(k)} = B^{(k-1)} - q_k q_k^T, \quad \text{for} \quad k = 1, \dots, n-1. \tag{3}$$

For the first term, $y_1 = a_1$ and $q_1 = a_1 / \left(a_1^T a_1 \right)^{1/2}$. For $k = 1, \dots, n-1$, from (3) we have

$$B^{(k)} = I_m - \sum_{i=1}^{k} q_i q_i^T. \tag{4}$$

Thus (2) gives $y_k = a_k - \sum_{i=1}^{k-1} \left(q_i^T a_k \right) q_i$. Identifying a_k with x_k and q_i with z_i, this is exactly the Gram-Schmidt process given by (1) as claimed.

A column of matrix A can be written as the sum of two orthogonal vectors, namely

$$a_k = B^{(k-1)} a_k + \sum_{i=1}^{k-1} \left(q_i^T a_k \right) q_i.$$

The first vector $B^{(k-1)} a_k$ is orthogonal to Span $\{q_1, \dots, q_{k-1}\}$, whereas the second vector (involving the summation) is in Span $\{q_1, \dots, q_{k-1}\}$ and is the projection of a_k on this subspace.

The matrices $B^{(k)}$ have some interesting basic properties which depend only on the fact that they are defined from an orthonormal sequence. We now summarize them (in Theorems 1 and 2) and then (in Theorem 3) prove a result that shows explicitly how zero rows can occur in $B^{(k)}$ and so force zero entries in the output of Gram-Schmidt.

THEOREM 1.

(i) *Given* $k \, (0 \le k \le n-1)$, *the matrix* $B^{(k)}$ *is positive semidefinite with rank* $m - k$.

(ii) *If* $0 \le i \le j \le n-1$, *then* $B^{(i)} \ge B^{(j)}$, *that is* $B^{(i)} - B^{(j)}$ *is positive semidefinite.*

(iii) *If* $0 \le i, j \le n-1$, *then* $B^{(i)} B^{(j)} = B^{(j)} B^{(i)} = B^{(q)}$ *where* $q = \max \{i, j\}$.

Proof.

(i) The fact that $B^{(k)}$ is symmetric is obvious from (4). As $q_i^T q_i = 1$ and $q_j^T q_i = 0$ for $i \ne j$, then $(B^{(k)})^2 = B^{(k)}$; that is $B^{(k)}$ is idempotent.

Thus the eigenvalues of $B^{(k)} \in \{0, 1\}$, and $B^{(k)}$ is positive semidefinite. (Note that $B^{(0)} = I_m$ is, in fact, positive definite.) Each product $q_i q_i^T$ is symmetric and has rank 1, and $q_i q_i^T q_j q_j^T = q_j q_j^T q_i q_i^T = 0$ for all $i \ne j$; so $\{q_i q_i^T\}, i = 1, \dots, n$, form a commuting family of symmetric matrices. Thus there exists a single orthogonal matrix U such that $U \left(q_i q_i^T \right) U^T = D_i$, where D_i is an m-by-m diagonal matrix for $i = 1, \dots, n$; see e.g., [4, 2.5.5]. Each D_i has rank 1, so it has exactly one nonzero entry (equal to 1), and if $i \ne j$, then $D_i D_j = U q_i q_i^T q_j q_j^T U^T = 0$. From (4) we see that $U B^{(k)} U^T = I_m - \sum_{i=1}^{k} D_i$, and thus rank $B^{(k)} = m - k$.

(ii) This follows directly from (3). Note that $B^{(i)} \ne B^{(j)}$ for $i \ne j$.

(iii) Suppose $q = i > j$, and consider

$$
\begin{aligned}
B^{(i)}B^{(j)} &= \left(B^{(j)} - q_{j+1}q_{j+1}^T - \cdots - q_i q_i^T\right) B^{(j)} \\
&= B^{(j)} - \left(q_{j+1}q_{j+1}^T + \cdots + q_i q_i^T\right)\left(I_m - q_1 q_1^T - \cdots - q_j q_j^T\right) \\
&= B^{(j)} - \left(q_{j+1}q_{j+1}^T + \cdots + q_i q_i^T\right) = B^{(i)} .
\end{aligned}
$$

The other products follow in a similar manner. $\qquad\square$

As $B^{(k)}$ is symmetric and idempotent, it follows that it is a projection matrix; hence we call $B^{(k)}$ a *Gram-Schmidt projection*. If the range of $B^{(k)}$ is S_k, then $B^{(k)}$ is the orthogonal projection onto S_k. We can deduce more about the spectrum of $B^{(k)}$ by using the rank result in Theorem 1 (i). In fact, $B^{(k)}$ has eigenvalue 1 with multiplicity $m - k$, and eigenvalue 0 with multiplicity k. The spectral properties show immediately that rank $B^{(k)} = $ trace $B^{(k)}$. From Theorem 1 (iii) and also the fact that $B^{(i)}B^{(j)}B^{(i)} = B^{(i)}$ for $i > j$, we note that $B^{(j)}$ is a $\{1,3,4,5\}$ generalized inverse of $B^{(i)}$; see, e.g., [1]. Also $B^{(k)}$ is its own $\{1,2,3,4,5\}$ inverse.

We use the properties of $\{B^{(k)}\}$ to deduce the following.

THEOREM 2. *If a sequence of m-by-m matrices $B^{(k)}, k = 0,\dots, n-1$, have the properties of Theorem 1 (i), (ii) and (iii), then they must satisfy (iii) for some orthonormal set of vectors $\{q_k\}$.*

Proof. First note that $C^{(k)} \equiv B^{(k-1)} - B^{(k)} \geq 0$, from Theorem 1 (ii), so if $x^T B^{(k-1)}x = 0$ then $x^T B^{(k)}x = 0$. With S_k denoting the subspace onto which $B^{(k)}$ projects, we have $S_k \subset S_{k-1}$. Letting $S_k^\perp$ denote the orthogonal complement of S_k, if $x \in S_k \cup S_{k-1}^\perp$ then $C^{(k)}x = 0$ (as $B^{(k-1)}x = B^{(k)}x = 0$ if $x \in S_{k-1}^\perp$; and $B^{(k)}x = x = B^{(k-1)}x$ if $x \in S_k$). Thus, by Theorem 1 (i), $C^{(k)}$ is not the zero matrix, so the dimension of the null space of $C^{(k)}$ is exactly $m - 1$, which implies that $C^{(k)}$ has rank one. We can therefore write $C^{(k)} = f_k f_k^T$ for some vector $f_k \in \Re^m$. As trace $C^{(k)} = 1$, we must have $f_k^T f_k = 1$. Also, by Theorem 1 (iii), $C^{(i)}C^{(j)} = 0$ for $i \neq j$, so $f_i f_i^T f_j f_j^T = \left(f_i^T f_j\right) f_i f_j^T = 0$, which implies $f_i^T f_j = 0$, and $\{f_k\}$ is an orthonormal set of vectors which we can identify with $\{q_k\}$. $\qquad\square$

Given k $(2 \leq k \leq n)$, t $(1 \leq t \leq k-1)$ and indices $1 \leq j_1 < j_2 < \cdots < j_t \leq k-1$, note that $B^{(r)}$, $r \geq k-1$, projects into the orthogonal complement of Span$\left\{a_{j_1},\dots, a_{j_t}\right\}$, since it projects onto the orthogonal complement of Span$\{a_1,\dots, a_r\}$. In particular, Span$\left\{a_{j_1},\dots, a_{j_t}\right\}$ is in the null space of $B^{(r)}$. In the event that $a_{j_1},\dots, a_{j_t}$ collectively have nonzero entries in *only* the t rows $i_1,\dots, i_t$, then Span$\left\{a_{j_1},\dots, a_{j_t}\right\}$ is exactly the coordinate subspace of $\Re^m$ in which there are arbitrary entries in positions $i_1,\dots, i_t$ and 0 entries elsewhere. For $B^{(r)}$ to have such a subspace in its null space, each row of $B^{(r)}$ must have zeros in the positions $i_1,\dots, i_t$. We have thus proved the following combinatorial result.

THEOREM 3. *If columns $a_{j_1},\dots, a_{j_t}$ for $1 \leq j_1 < \cdots < j_t \leq k-1$ and $2 \leq k \leq n$ collectively have nonzero entries only in rows $i_1,\dots, i_t$, then rows and columns $i_1,\dots, i_t$ of $B^{(r)}$ are zero for $k-1 \leq r \leq n-1$.* $\qquad\square$

It follows from (2) that if some subset of the columns of A has the property stated in Theorem 3, then the entries $i_1, \ldots, i_t$ of vectors $q_{r+1}, \ldots, q_n$ are zero.

The following example illustrates the construction and some properties of the Gram-Schmidt projections.

Consider $A = \begin{bmatrix} 1 & 3 & 1 \\ 0 & 0 & 1 \\ 2 & 4 & 2 \end{bmatrix}$. Then (2) and (3) give:

$$B^{(0)} = I_3, \qquad q_1 = \frac{1}{\sqrt{5}}(1, 0, 2)^T,$$

$$B^{(1)} = I_3 - q_1 q_1^T = \begin{bmatrix} \frac{4}{5} & 0 & -\frac{2}{5} \\ 0 & 1 & 0 \\ -\frac{2}{5} & 0 & \frac{1}{5} \end{bmatrix}, \qquad q_2 = \frac{1}{\sqrt{5}}(2, 0, -1)^T,$$

$$B^{(2)} = B^{(1)} - q_2 q_2^T = \begin{bmatrix} 0 & 0 & 0 \\ 0 & 1 & 0 \\ 0 & 0 & 0 \end{bmatrix}, \qquad q_3 = (0, , 1, 0)^T.$$

Here columns a_1, a_2 have nonzero entries only in rows 1,3; thus rows and columns 1 and 3 of $B^{(2)}$ are zero and the 1 and 3 entries of q_3 are zero. $\qquad\Box$

References

1. Ben-Israel, A. and Greville, T.N.E., *Generalized Inverses: Theory and Applications*, Wiley, 1974.

2. Golub, G.H. and Van Loan, C.F., *Matrix Computations*, Johns Hopkins University Press, 1989.

3. Hare, D.R., Johnson, C.R., Olesky, D.D., and van den Driessche, P., Sparsity Analysis of the QR Factorization, *SIAM J. Matrix Anal. Appl.* **14** (1993), 655-669.

4. Horn, R.A. and Johnson, C.R., *Matrix Analysis*, Cambridge University Press, 1985.

Charles R. Johnson
The College of William and Mary

D.D. Olesky and P. van den Driessche
University of Victoria, British Columbia
Contributed

Pythagoras and the Cauchy-Schwarz Inequality

Most current textbook introductions to coordinate vector algebra appeal to the students' knowledge of Euclidean geometry and geometric intuition to justify the introduction of Cartesian coordinates and the usual distance formula. Similarly, vector addition, scalar multiplication, and length of a vector are accompanied by discussions of their geometric significance. Frequently, however, the usual inner (dot) product then appears abruptly, the Cauchy-Schwarz and triangle inequalities are proved as technical exercises, and the cosine of the angle between vectors is defined with no reference to the ratio of sides of a right triangle. The following remarks indicate one way that elementary geometric considerations lead naturally to the introduction of the inner product, to trivial proofs of the inequalities, and to the geometric law of cosines.

We assume that the distance between points (vectors) A and B in R^n and the length of $B - A$ may be computed as $|B - A| = \left(\sum (b_i - a_i)^2\right)^{1/2}$, and this has been justified by appeal to the Pythagorean Theorem. Considering the triangle with vertices $A, B, -A$ it is clear the vector B should be *orthogonal* to vector A if and only if $|B - A| = |B - (-A)|$, so we take this as our definition of orthogonality.

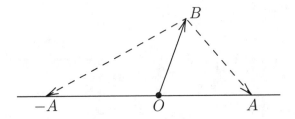

Since

$$|B - A|^2 = |A|^2 + |B|^2 - 2\sum a_i b_i \quad \text{and} \quad |B + A|^2 = |A|^2 + |B|^2 + 2\sum a_i b_i \ ,$$

B is orthogonal to A if and only if $\sum a_i b_i = 0$ and also if and only if the Pythagorean identity $|B - A|^2 = |A|^2 + |B|^2$ holds. Moreover, when A and B are not orthogonal, the quantity $\sum a_i b_i$ measures the failure of the Pythagorean identity. Now define the *dot (scalar, inner) product* by $A \cdot B = \sum a_i b_i$ and note that, like ordinary number multiplication, it is bilinear and positive.

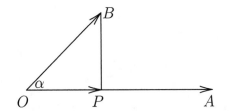 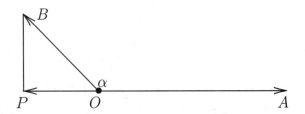

The point P is the *orthogonal projection of B onto A* (onto the line thru the origin O and A) if and only if $P = cA$ for some number c and $B - P$ is orthogonal to A. From $A \cdot (B - cA) = 0$ it follows that $c = A \cdot B / A \cdot A$ and, as we saw above, since $B - P$ is orthogonal to P it follows that $|B|^2 = |P|^2 + |B - P|^2$. Hence

$$1 \geq \frac{|P|}{|B|} = \frac{|A \cdot B|}{|A||B|} \qquad \text{(\textit{Cauchy-Schwarz inequality})}$$

and equality holds if $B = P$, that is, $B = cA$. Moreover, in the right triangle O, P, B the ratio $|P|/|B|$ is the usual cosine of the angle at O. Taking into account the sign of $A \cdot B$, it is geometrically clear that the angle α between A and B satisfies

$$\cos \alpha = \frac{A \cdot B}{|A||B|} \, .$$

From the computation of $|B - A|^2$ above we get $|B - A|^2 = |A|^2 + |B|^2 - 2|A||B| \cos \alpha$ (*Law of Cosines*). This equation also makes obvious the *triangle inequality*

$$|A|^2 + |B|^2 - 2|A||B| \leq |B - A|^2 \leq |A|^2 + |B|^2 + 2|A||B| \text{ or}$$

$$\left| |A| - |B| \right| \leq |B - A| \leq |A| + |B|.$$

Finally note that the arguments above work equally well in any space with an inner product.

Ladnor Geissinger
University of North Carolina, Chapel Hill
American Mathematical Monthly **83** (1976), 40–41

An Application of the Schwarz Inequality

In textbooks on linear algebra a complex $n \times n$ matrix U is said to be *unitary* if

$$U^*U = I_n$$

where U^* is the transposed conjugate of U. The following theorem is then given:

Theorem. *For the matrix U to be unitary it is necessary and sufficient that $\|Ux\| = \|x\|$, for all x in unitary n-space, $\mathcal{U}_n$.*

Although the necessity is trivial to prove, the sufficiency part is a bit more involved. A variety of proofs, for the most part being variations of those in [1] and [2], are in the literature. None of the proofs which we have seen make use of the Schwarz inequality. In this note we show how the Schwarz inequality can be employed to construct a brief proof of the sufficiency part of the above theorem. This new proof can be used as an application of the Schwarz inequality which usually has been presented earlier in the book where its wide applicability has been duly noted but rarely if ever illustrated.

Thus we assume that

$$(1) \qquad \|Ux\|^2 = (Ux, Ux) = \|x\|^2 = (x, x), \qquad \text{for all } x \in \mathcal{U}_n,$$

and we prove that U is a unitary matrix. We first note that (1) implies that U, hence also U^*, is nonsingular. Now suppose that x is an arbitrary nonzero vector from $\mathcal{U}_n$. Then

$$
\begin{aligned}
(U^*x, U^*x)^2 \; &= (x, UU^*x)^2 \le (x, x)(UU^*x, UU^*x) \\
&= (x, x)(U^*x, U^*x),
\end{aligned}
$$

where we have used the Schwarz inequality and (1). Then, since $(U^*x, U^*x) > 0$,

$$(2) \qquad (U^*x, U^*x) \le (x, x).$$

Also,

$$
\begin{aligned}
(3) \qquad (x, x)^2 \; &= (Ux, Ux)^2 = (x, U^*Ux)^2 \le (x, x)(U^*Ux, U^*Ux) \\
&\le (x, x)(Ux, Ux) \\
&= (x, x)^2
\end{aligned}
$$

where we have used first the Schwarz inequality and then (2). Thus equality must hold in the Schwarz inequality, which means that $U^*Ux = \alpha x$ for some scalar α. But

$$\alpha = \frac{(U^*Ux, x)}{(x, x)} = \frac{(Ux, Ux)}{(x, x)} = 1$$

105

so that $U^*Ux = x$ for all nonzero $x \in \mathcal{U}_n$ (equality is trivially true for $x = 0$), or

$$U^*U = I_n.$$

As an alternative way of finishing the proof we note that (3) implies equality in (2) and this with (1) implies that

$$(x - U^*Ux, x - U^*Ux) = 0$$

or $x - U^*Ux = 0$ for all $x \in \mathcal{U}_n$.

References

1. F.E. Hoh, *Elementary Matrix Algebra*, 2nd ed., Macmillan, New York, 1964.
2. P.R. Halmos, *Finite Dimensional Vector Spaces*, Van Nostrand, Princeton, NJ, 1958.

V.G. Sigillito
The Johns Hopkins University
American Mathematical Monthly **75** (1968), 656–658

A Geometric Interpretation of Cramer's Rule

If A is a nonsingular real $n \times n$ matrix and $b \in \Re^n$, then Cramer's rule states that if $Ax = b$, then the components x_i of x satisfy

$$(1) \qquad x_i = \frac{\det A_i}{\det A},$$

in which A_i is the matrix obtained from A by replacing the ith column of A by b.

One may interpret (1) as expressing x_i as the ratio of the (signed) volumes of two n-dimensional parallelepipeds. Thus, it might be interesting to see a proof of (1) based upon geometric principles.

Our proof of Cramer's rule will be for the 3×3 case, and for the sake of exposition we will make certain simplifying assumptions. First, we will assume that A has a positive determinant, and second that our right-hand side b lies in the open cone generated by the columns of A (that is, the set of linear combinations of columns of A in which the coefficients are positive).

Let A be a nonsingular 3×3 matrix, let $a_1, a_2,$ and a_3 be the columns of A, and let $b \in \Re^3$. Since A is nonsingular, there exist uniquely determined scalars x_1, x_2 and x_3 such that

$$x_1 a_1 + x_2 a_2 + x_3 a_3 = b.$$

(From our simplifying assumptions x_1, x_2 and x_3 are positive.)

Consider the two parallelepipeds generated by the vectors $x_1 a_1, x_2 a_2$ and $x_3 a_3$ and by b, $x_2 a_2$ and $x_3 a_3$, as pictured in the figure below.

Think of the parallelogram generated by $x_2 a_2$ and $x_3 a_3$ as the base of both of these parallelepipeds. Note that these two parallelepipeds then have the same perpendicular height above the base (indeed, the face opposite the common base lies, in both cases, in the plane $x_1 a_1 + \text{Span}\,\{a_2, a_3\}$). It follows that our two parallelepipeds have equal volumes. Thus

$$\det \begin{bmatrix} x_1 a_1 & x_2 a_2 & x_3 a_3 \end{bmatrix} = \det \begin{bmatrix} b & x_2 a_2 & x_3 a_3 \end{bmatrix}.$$

It follows then that

$$x_1 x_2 x_3 \det \begin{bmatrix} a_1 & a_2 & a_3 \end{bmatrix} = x_2 x_3 \det \begin{bmatrix} b & a_2 & a_3 \end{bmatrix},$$

and equation (1) follows for the case $i = 1$; of course, the cases $i = 2$ and 3 are entirely analogous.

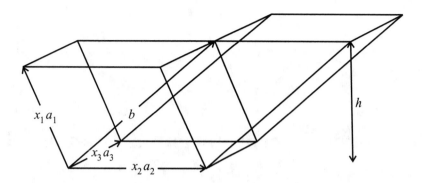

Gregory Conner and Michael Lundquist
Brigham Young University
Contributed

Isometries of l_p-norm

The goal of this note is to provide a short proof of the fact that the isometries of l_p-norm ($p \neq 2$) on $\mathbf{R}^n$ are generalized permutation matrices, those matrices that can be written as a product of a diagonal orthogonal matrix and a permutation matrix. As usual, the l_p=norm on $\mathbf{R}^n$ is defined by $l_p(x) = \left(\sum_{i=1}^n |x_i|^p\right)^{1/p}$ if $1 \leq p < \infty$, and $l_p(x) = \max_{1 \leq i \leq n} |x_i|$ if $p = \infty$. An isometry of l_p-norm is an $n \times n$ matrix A satisfying

$$l_p(Ax) = l_p(x) \quad \text{for all} \quad x \in \mathbf{R}^n.$$

It is well-known that the isometries of the l_2-norm are orthogonal matrices. A less well-known fact is the following theorem.

Theorem. *Suppose $1 \leq p \leq \infty$ and $p \neq 2$. An $n \times n$ matrix A is an isometry of l_p-norm if and only if A is a generalized permutation matrix.*

This result can be deduced from stronger statements about more general norms [2][3][7] or can be viewed as a special case of its infinite dimensional version [1,p. 119][5,p. 112]. However, there is a direct proof which requires only a basic fact from the theory of norm on $\mathbf{R}^n$ [4,Ch. 5]: an $n \times n$ matrix A is an isometry of l_p-norm if and only if its transpose A^T is an isometry of l_q-norm, where $1/p + 1/q = 1$.

Proof. The sufficiency is easy to check. For necessity, suppose $A = (a_{ij})$ is an isometry of the l_p-norm. We may assume $1 \leq p < 2$, otherwise consider the l_q-norm and A^T. First consider the case that $p \neq 1$. For $i = 1, \dots, n$ let e_i denote the ith column of the $n \times n$ identity matrix. Then $l_p(Ae_i) = 1$ for all $i = 1, \dots, n$. Thus $|a_{ij}| \leq 1$ and $\sum_{i,j=1}^n |a_{ij}|^p = n$. Since A^T is an isometry of the l_q-norm, the same argument gives $\sum_{i,j=1}^n |a_{ij}|^q = n$. Notice that $|a_{ij}|^q \leq |a_{ij}|^p$, and equality holds if and only if $|a_{ij}| = 0$ or 1. Since every column of A (respectively, A^T) has l_p-norm (respectively, l_q-norm) equal to one, the equality $\sum_{i,j=1}^n |a_{ij}|^p = n = \sum_{i,j=1}^n |a_{ij}|^q$ implies that each row and each column of A has exactly one nonzero entry whose magnitude equals one, and the result follows. For the case $p = 1$, the above argument also yields $\sum_{i,j=1}^n |a_{ij}| = n = \sum_{i=1}^n \alpha_i = \max_{1 \leq j \leq n} |a_{ij}|$. Thus each row and hence each column of A has only one nonzero entry with magnitude equal to one, and the result follows. ■

After this note was finished, the authors were informed that R. Mathias [6] had obtained the same proof for the complex case previously and independently.

ACKNOWLEDGEMENT. The authors wish to thank Professor R. Horn for bringing the references [6] [7] to their attention.

References

1. B. Beauzamy, *Introduction to Banach Spaces and Their Geometry*, North Holland, 1985.

2. S. Chang and C.K. Li, Certain isometries on $\mathbf{R}^n$, *Linear Algebra and Its Applications* **165** (1992), 251–265.

3. D.Z. Dokovic, C.K. Li and L. Rodman, Isometries of symmetric gauge functions, *Linear and Multilinear Algebra* **30** (1991), 81–92.

4. R. Horn and C.R. Johnson, *Matrix Analysis*, Cambridge University Press, 1985.

5. J. Linderstrauss and L. Tzafriri, *Classical Banach Spaces I: Sequence Spaces*, Springer-Verlag, 1977.

6. R. Mathias, unpublished note.

7. C. Wang et. al, Structure of p-isometric matrices and rectangular matrices with minimum p-norm condition number, *Linear Algebra and Its Applications* **184** (1993), 261-278.

Chi-Kwong Li
College of William and Mary

Wasin So
Sam Houston State University
American Mathematical Monthly **101** (1994), 452–453

Matrices Which Take a Given Vector into a Given Vector

A slight variation of the usual problem for a system of linear equations is to solve for the coefficient matrix when the solution vector is known. More abstractly, if V and W are n- and m-dimensional vector spaces over the same field, and $x \in V$ and $y \in W$, find all linear transformations L such that $Lx = y$. In this note we answer the question completely and in fact give a procedure for writing down the solution from inspection.

Since we are dealing with the finite-dimensional case, we may assume x and y are given column vectors and L is an $m \times n$ matrix. When $x = 0$, a solution clearly exists if and only if $y = 0$, and L is, in fact, any $m \times n$ matrix. The case for $x \neq 0$ is governed by the following:

THEOREM. *Let $x \neq 0$ and y be column vectors of n and m components, respectively (with elements from the same field). Suppose the p-th component of x is not zero. The set of all matrices L such that $Lx = y$ is nonempty, and is given by*

$$L = L^0 + \sideset{}{'}\sum_{r=1}^{n} \sum_{q=1}^{m} \alpha_{qr} L^{qr} \tag{1}$$

(the prime denotes that r does not take on the value p), where the α_{qr} are arbitrary constants (scalars of the field), and

$$L_{ij}^0 = (y_i \delta_{jp})/x_p, \qquad i = 1, 2, \cdots, m; \quad j = 1, 2, \cdots, n,$$

(2) $\qquad\qquad\qquad\qquad\qquad$ *(δ_{ij} is the "Kronecker delta")*

$$L_{ij}^{qr} = x_r \delta_{iq} \delta_{jp} - x_p \delta_{iq} \delta_{jr}, \quad i = 1, 2, \cdots, m; \quad j = 1, 2, \cdots, n.$$

Proof. The equation $Lx = y$ or

(3)
$$\begin{aligned}
L_{11}x_1 + L_{12}x_2 + \cdots + L_{1n}x_n &= y_1 \\
L_{21}x_1 + L_{22}x_2 + \cdots + L_{2n}x_n &= y_2 \\
&\cdots\cdots\cdots\cdots\cdots\cdots \\
L_{m1}x_1 + L_{m2}x_2 + \cdots + L_{mn}x_n &= y_m
\end{aligned}$$

may be rewritten as

$$(4) \quad \begin{bmatrix} x_1 & x_2 & \cdots & x_n & 0 & 0 & \cdots & 0 & 0 & \cdots & 0 & \cdots\cdots\cdots \\ 0 & 0 & \cdots & 0 & x_1 & x_2 & \cdots & x_n & 0 & \cdots & 0 & \cdots\cdots\cdots \\ \cdots\cdots\cdots\cdots\cdots\cdots\cdots\cdots\cdots\cdots\cdots\cdots\cdots\cdots\cdots\cdots \\ \cdots\cdots\cdots\cdots\cdots\cdots\cdots\cdots\cdots\cdots\cdots\cdots\cdots\cdots\cdots\cdots \\ 0 & 0 & \cdots & 0 & 0 & 0 & \cdots & 0 & 0 & \cdots & 0 & x_1 x_2 \cdots x_n \end{bmatrix} \begin{bmatrix} L_{11} \\ L_{12} \\ \vdots \\ L_{1n} \\ L_{21} \\ L_{22} \\ \vdots \\ L_{2n} \\ \vdots \\ L_{mn} \end{bmatrix} = \begin{bmatrix} y_1 \\ y_2 \\ \vdots \\ y_m \end{bmatrix}$$

Denoting this $m \times mn$ coefficient matrix by X, we see that its rank is m since $x_p \neq 0$. Since the rank of the augmented matrix is also m there is a solution L to this system. Also the associated homogeneous system has $nm - m = (n-1)m$ linearly independent solutions. Thus the general solution is the sum of a particular solution plus an arbitrary linear combination of $(n-1)m$ linearly independent solutions of the homogeneous system. One may readily check that L^0 and the L^{qr} as defined by (2) are these required solutions. (They may be obtained by applying Gaussian elimination to (4).)

We observe that L^0 and the L^{qr} may be immediately written down from an inspection of (3). The pth column of L^0 is y/x_p, all the other entries are zero. The qth row of L^{qr} contains x_r in the pth column and $-x_p$ in the rth column. All its other rows are zero. Thus the set of all matrices A satisfying

$$\begin{bmatrix} a_{11} & a_{12} & a_{13} \\ a_{21} & a_{22} & a_{23} \end{bmatrix} \begin{bmatrix} 1 \\ -1 \\ 2 \end{bmatrix} = \begin{bmatrix} 1 \\ 5 \end{bmatrix}$$

is immediately seen to be (we will work with $p = 1$)

$$A = \begin{bmatrix} 1 & 0 & 0 \\ 5 & 0 & 0 \end{bmatrix} + \alpha_{12} \begin{bmatrix} -1 & -1 & 0 \\ 0 & 0 & 0 \end{bmatrix} + \alpha_{22} \begin{bmatrix} 0 & 0 & 0 \\ -1 & -1 & 0 \end{bmatrix}$$

$$+ \alpha_{13} \begin{bmatrix} 2 & 0 & -1 \\ 0 & 0 & 0 \end{bmatrix} + \alpha_{23} \begin{bmatrix} 0 & 0 & 0 \\ 2 & 0 & -1 \end{bmatrix}$$

Although the question is easily answered in the finite-dimensional case it offers new difficulties when we proceed to infinite dimensions, especially when we restrict the class of permissible transformations. A typical question might be to find all kernels of the integral equation $\int_a^b k(s,t)f(t)dt = g(s)$ once f and g are specified.

M. Machover
St. John's University
American Mathematical Monthly **74** (1967), 851–852

The Matrix of a Rotation

What is the matrix of the rotation of R^3 about a unit axis $\mathbf{p}$ through an angle θ? Since a rotation has a fixed axis (or eigenvector or eigenvalue 1) and rotates the plane perpendicular to $\mathbf{p}$ by angle θ, the matrix is easy to determine if we change to a convenient basis; however, it is not well known that there is a simple expression for the matrix in the standard basis which depends only on the coordinates of $\mathbf{p}$ and the angle θ. The formula is obtained without changing bases. Furthermore, this formula can be useful in coding the effect of a rotation in computer graphics. The derivation of this formula was motivated by the close relationship between rotations and quaternions.

For two vectors $\mathbf{v}$ and $\mathbf{w}$ we use the notation $\mathbf{v} \cdot \mathbf{w}$ for the standard inner product and $\mathbf{v} \times \mathbf{w}$ for the cross product. Consider the linear transformations of R^3 given by $P(\mathbf{q}) = \mathbf{p} \times \mathbf{q}$ where $\mathbf{p}$ is a unit vector.

Proposition 1. $P^2(\mathbf{q}) = -\mathbf{q} + (\mathbf{p} \cdot \mathbf{q})\mathbf{p}$. *Thus $I + P^2$ is the projection operator along the unit vector $\mathbf{p}$.*

Proof. The first part follows easily by using the triple product formula $\mathbf{p} \times (\mathbf{q} \times \mathbf{r}) = (\mathbf{p} \cdot \mathbf{r})\mathbf{q} - (\mathbf{p} \cdot \mathbf{q})\mathbf{r}$. The second statement follows easily from the first part and the definition of an orthogonal projection.

Proposition 2. *The rotation of R^3 about an axis $\mathbf{p}$ of unit length by an angle θ is given by*

$$L(\mathbf{q}) = \mathbf{q} + (\sin\theta)P(\mathbf{q}) + (1 - \cos\theta)P^2(\mathbf{q}).$$

Proof. To see this, we show that this linear transformation has the geometric properties of a rotation as described in the first paragraph. The vector $\mathbf{p}$ is left fixed since $P(\mathbf{p}) = P^2(\mathbf{p}) = 0$. It follows from the definition of L and Proposition 1 that if $\mathbf{q}$ is perpendicular to $\mathbf{p}$ then $L(\mathbf{q}) = \cos\theta\mathbf{q} + \sin\theta(\mathbf{p} \times \mathbf{q})$. Moreover if $\mathbf{q}$ is also of unit length then $\mathbf{p} \times \mathbf{q}$ is of unit length and also perpendicular to $\mathbf{p}$; hence

$$L(\mathbf{p} \times \mathbf{q}) = \cos\theta(\mathbf{p} \times \mathbf{q}) + \sin\theta(\mathbf{p} \times (\mathbf{p} \times \mathbf{q})) = \cos\theta(\mathbf{p} \times \mathbf{q}) - \sin\theta\mathbf{q}.$$

Thus the plane perpendicular to $\mathbf{p}$ is rotated by angle θ. It follows now that all of R^3 is rotated about the axis $\mathbf{p}$ by angle θ and thus L describes the rotation.

We can now easily write the matrix of $L = I + (\sin\theta)P + (1 - \cos\theta)P^2$ in terms of the standard basis $\mathbf{e}_1, \mathbf{e}_2, \mathbf{e}_3$. Suppose $\mathbf{p} = (a, b, c)^t$; then P is easily computed: $P(\mathbf{e}_1) =$

$(0, c, -b)^t$, $P(\mathbf{e}_2) = (-c, 0, a)^t$ and $P(\mathbf{e}_3) = (b, -a, 0)^t$. Furthermore, the matrix of the projection $I + P^2$ is the matrix product $\mathbf{p}\mathbf{p}^t$. Thus the matrix of L is

$$I + (\sin\theta) \begin{bmatrix} 0 & -c & b \\ c & 0 & -a \\ -b & a & 0 \end{bmatrix} + (1 - \cos\theta) \begin{bmatrix} a^2 - 1 & ab & ac \\ ab & b^2 - 1 & bc \\ ac & bc & c^2 - 1 \end{bmatrix}.$$

Roger C. Alperin
San Jose State University
College Mathematics Journal **20** (1989), 230

Standard Least Squares Problem

Consider the equation

$$A\vec{x} = \vec{y} \tag{2}$$

where A is an $m \times n$ matrix with $m \geq n$, $\vec{x}$ is in R^n, and $\vec{y}$ is in R^m. The least squares solution to this equation is the vector $\vec{x}^*$ that satisfies the equation

$$A^T A \vec{x}^* = A^T \vec{y}, \tag{3}$$

which is called the normal equation. This result may be derived by the methods of calculus, geometry, or algebra.

Using calculus, one takes derivatives of the objective function

$$J(\vec{x}) = \|A\vec{x} - \vec{y}\|^2 \tag{4}$$

with respect to $x_1, x_2, \cdots, x_n$, the components of $\vec{x}$. Setting these derivatives equal to zero, solving the resulting equations, and rearranging yield equation (2).

The geometric approach recognizes that for any candidate solution $\vec{x}$, the vector $A\vec{x}$ will be in the column space of A. Hence, one seeks the vector in the column space of A that is closest to $\vec{y}$. This implies $\vec{x}^*$ satisfies

$$
\begin{aligned}
A\vec{x}^* &= \text{projection of } \vec{y} \text{ onto the column space of } A \\
&= \text{projection matrix of } A \text{ times } \vec{y} \\
&= A(A^T A)^{-1} A^T \vec{y}
\end{aligned}
$$

and

$$\vec{x}^* = (A^T A)^{-1} A^T \vec{y},$$

assuming the columns of A are linearly independent. This approach provides the most insight and is the approach taken in [1].

The algebraic approach takes the method of "completing the square" from high school algebra and extends it to vectors and matrices. As motivation, consider the quadratic polynomial $f(x) = ax^2 + 2bx + c$ where $a > 0$. The minimum can be found by adding and subtracting b^2/a to complete the square. Specifically,

$$\begin{aligned}
f(x) &= ax^2 + 2bx + c \\
&= a\left(x^2 + \frac{2b}{a}x\right) + c \\
&= a\left[x^2 + \frac{2b}{a}x + \left(\frac{b}{a}\right)^2\right] - \frac{b^2}{a} + c \\
&= a\left(x + \frac{b}{a}\right)^2 - \frac{b^2}{a} + c.
\end{aligned} \tag{5}$$

The minimizing value of x can now be found by inspection. The only term that depends on x is $a(x + (b/a))^2$, which is obviously minimized by $x = -b/a$. Therefore, the quadratic polynomial is minimized by $x = -b/a$. In anticipation of the vector/matrix generalization, this polynomial is rewritten as

$$f(x) = (x + a^{-1}b)a(x + a^{-1}b) - ba^{-1}b + c. \tag{6}$$

The objective function in equation (3) can be expressed as

$$\begin{aligned}
J(\vec{x}) &= (A\vec{x} - \vec{y})^T(A\vec{x} - \vec{y}) \\
&= \vec{x}^T A^T A\vec{x} - \vec{x}^T A^T \vec{y} - \vec{y}^T A\vec{x} + \vec{y}^T \vec{y} \tag{7} \\
&= \vec{x}^T S\vec{x} - 2\vec{x}^T \vec{z} + \vec{y}^T \vec{y} \tag{8}
\end{aligned}$$

where

$$\begin{aligned}
S &= A^T A, \\
\vec{z} &= A^T \vec{y}.
\end{aligned}$$

This is analogous to the quadratic polynomial as expressed in equation (4) above with

$$\begin{aligned}
x &\leftrightarrow \vec{x} \\
a &\leftrightarrow S \\
b &\leftrightarrow -\vec{z} \\
c &\leftrightarrow \vec{y}^T \vec{y}.
\end{aligned}$$

If the columns of A are linearly independent, then $S = A^T A$ is invertible. (See [1] Theorem 19, p. 194.) In addition, A is symmetric ($S^T = S$). These properties allow the objective function to be rewritten as

$$J(\vec{x}) = (\vec{x} - S^{-1}\vec{z})^T S(\vec{x} - S^{-1}\vec{z}) - \vec{z}^T S^{-1}\vec{z} + \vec{y}^T \vec{y}, \tag{9}$$

which is analogous to equation (5); see Exercise (4) below. S has the property $\vec{v}^T S\vec{v} > 0$ if $\vec{v} \neq 0$ by Exercises (1) and (2). (Matrices with this property are called positive definite.)

Hence, the unique value of $\vec{x}$ that minimizes $J(\vec{x})$ can be determined by inspection. It is $\vec{x}^* = S^{-1}\vec{z}$, which is equation (2).

If the columns of A are linearly independent, a similar argument also shows that $J(\vec{x})$ is minimized by any vector satisfying equation (2). In this case, however, $\vec{x}^*$ is not unique.

The simplicity and elegance of this solution reflects the compatibility of linear "dynamics"–equation (1) and a quadratic cost function–equation (3). It has been said that they fit together like a hand fits a glove.

Exercises

1. Show that $\vec{v}^T S \vec{v} \geq 0$ for all $\vec{v}$, where $S = A^T A$.

2. Show that $\vec{v}^T S \vec{v} \neq 0$ if $\vec{v} \neq 0$ (assuming the columns of A are linearly independent).

3. Show that $\vec{x}^T A^T \vec{y} = \vec{y}^T A \vec{x}$. (This is used in going from equation (6) to (7). The proof should be *very* short.)

4. Show that the form of $J(\vec{x})$ in equation (8) is equal to that given in equation (7).

References

1. L.W. Johnson, R.D. Riess, J.T. Arnold, *Introduction to Linear Algebra*, 3rd ed., Addison Wesley, Reading, MA.

James Foster
Santa Clara University
Contributed

Multigrid Graph Paper

Ordinary two-dimensional graph paper is imprinted with two sets of parallel lines that are mutually perpendicular to one another. Such graph paper is very useful for plotting points that are represented with respect to the canonical basis, $e_1 = (1,0)^T, e_2 = (0,1)^T$, of the vector space $\mathbf{R}^2$. Multigrid graph paper may be imprinted with three or four sets of parallel lines as shown in Figure 1. Such graph paper is easily produced using most of the drawing programs available for personal computers. It is also far more useful than ordinary graph paper for illustrating many concepts of elementary linear algebra, such as representation of vectors or linear transformations, change of basis, similarity, orthogonality, projections, reflections, and quadratic forms.

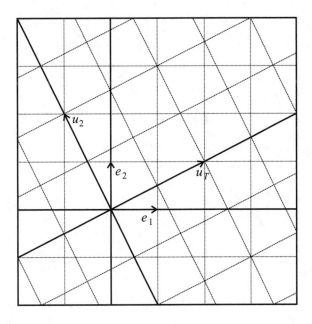

Figure 1

Consider the problem of writing the vector $v = (3,4)^T$ as a linear combination of the vectors $u_1 = (2,1)^T$ and $u_2 = (-1,2)^T$. Instead of solving a system of linear equations, one may simply write down the answer $v = 2u_1 + u_2$ by plotting the point $(3,4)$ on the e_1, e_2-grid in Figure 1, and then determining the coordinates of that point with respect to the u_1, u_2-grid.

119

This ability is particularly useful for working examples in $\mathbf{R}^2$ that would otherwise require solving several systems of equations. For example, if one has two bases for $\mathbf{R}^2$ and needs to express each vector in terms of the basis to which it does not belong, then one would have four systems of equations to solve. In such situations, students can become bogged down in solving the systems of equations, and never get to the underlying concept that has to be illustrated. Thus multigrid graph paper not only provides geometric interpretations for many concepts, but also allows students to work examples in which their efforts are more highly focused.

When one writes a vector v as a linear combination, $v = \alpha u_1 + \beta u_2$, of a basis $B = \{u_1, u_2\}$, the coefficients α, β are called the *coordinates* of v with respect to the basis B, and B is said to define a *coordinate system*. Without multigrid graph paper, students never *see* any coordinate system except the natural one associated with the canonical basis of $\mathbf{R}^2$. For a deeper understanding of coordinate systems students need to see other examples. In order to appreciate the properties of a coordinate system associated with an orthonormal basis, one would need to do some work with a nonorthonormal coordinate system such as the one associated with the basis $\{v_1, v_2\}$ of Figure 2.

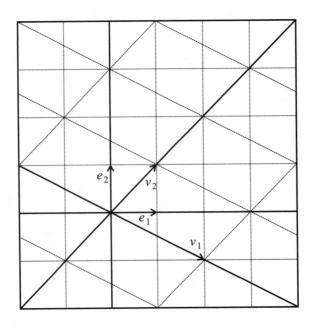

Figure 2

A change of basis in the representation of a vector or linear transformation is actually a change in coordinate system, and multigrid graph paper is particularly useful in this context. We illustrate this with the bases $B_0 = \{e_1, e_2\}$ and $B_1 = \{v_1, v_2\}$ of Figure 2. If w_0 gives the coordinates of some vector with respect to the basis B_0, and w_1 gives the coordinates of the same vector with respect to B_1, then there are invertible matrices S and T such that $w_1 = Sw_0$ and $w_0 = Tw_1$. The basic rule for obtaining a change of basis matrix states that the columns of S are the coordinates of e_1 and e_2 with respect to B_1, and the columns of T are the coordinates of v_1 and v_2 with respect to B_0. One immediately obtains $T = \begin{pmatrix} 2 & 1 \\ -1 & 1 \end{pmatrix}$. Also by examining Figure 2, $3e_1 = v_1 + v_2$ and $3e_2 = -v_1 + 2v_2$. (Note that $\det(T) = 3$, which is the area of the parallelogram with vertices at the origin, $v_1, v_1 + v_2$, and v_2.) Hence

$$e_1 = \frac{1}{3}v_1 + \frac{1}{3}v_2, \; e_2 = -\frac{1}{3}v_1 + \frac{2}{3}v_2, \; \text{and} \; S = \begin{pmatrix} 1/3 & -1/3 \\ 1/3 & 2/3 \end{pmatrix}.$$

It is now easy to check that $S = T^{-1}$. There is a pedagogical advantage in using just one rule to obtain both S and T rather than obtaining S as T^{-1}. To illustrate a basic property of a change of basis matrix, one may compute or check visually, for a point such as $(4, 1)$, that $S^{-1}\begin{pmatrix} 1 \\ 2 \end{pmatrix} = \begin{pmatrix} 4 \\ 1 \end{pmatrix}$ and $S\begin{pmatrix} 4 \\ 1 \end{pmatrix} = \begin{pmatrix} 1 \\ 2 \end{pmatrix}$ give the appropriate change of coordinates. One may similarly find a change of basis matrix for converting between B_0 and the basis $B_2 = \{u_1, u_2\}$ of Figure 1. By combining this matrix with S^{-1}, one can convert between coordinates with respect to B_1 and B_2. As illustrated above, multigrid graph paper not only simplifies the computation of change of basis matrices, but also provides a geometric setting for illustrating many of their properties.

Let $B_2 = \{u_1, u_2\}$ be the basis of Figure 1, and consider the linear transformation $L : \mathbf{R}^2 \to \mathbf{R}^2$ which doubles the component of a vector along u_1 and reverses the component along u_2; that is, L is defined by $L(u_1) = 2u_1$ and $L(u_2) = -u_2$. The columns of a matrix representation of L are the coordinates of the images under L of the basis elements. Thus, the matrix representation of L with respect to the basis B_2 is the matrix $B = \begin{pmatrix} 2 & 0 \\ 0 & -1 \end{pmatrix}$. From Figure 1 (expanded slightly to the right and upward), $5e_1 = 2u_1 - u_2$ and $5e_2 = u_1 + 2u_2$. Hence,

$$L(e_1) = \frac{4}{5}u_1 + \frac{1}{5}u_2 = \begin{pmatrix} 7/5 \\ 6/5 \end{pmatrix}, \quad L(e_2) = \frac{2}{5}u_1 - \frac{2}{5}u_2 = \begin{pmatrix} 6/5 \\ -2/5 \end{pmatrix},$$

and $A = \begin{pmatrix} 7/5 & 6/5 \\ 6/5 & -2/5 \end{pmatrix}$ is the matrix representation of L with respect to the basis B_0. One may now illustrate by computation or geometrically a number of properties such as

(1) $S^{-1}AS = B$ for the appropriate change of basis matrix S, so that A or B are similar.

(2) $A\begin{pmatrix} 1 \\ 3 \end{pmatrix} = \begin{pmatrix} 5 \\ 0 \end{pmatrix}$ does give $L\left(\begin{pmatrix} 1 \\ 3 \end{pmatrix}\right)$.

(3) u_1 and u_2 are eigenvectors of A (and L) corresponding to eigenvalues 2 and -1.

Similar treatments can be given to linear transformations defined as reflections about u_1 or u_2, or as projections onto u_1 or u_2. Such examples reinforce geometric ideas related to linear transformations.

Multigrids should have two or three times more lines than shown in each parallel set of Figures 1 and 2. This would prevent having to extend the grid as we did in the last example. I like to use a full $8\frac{1}{2}$ by 11-inch sheet for each multigrid, and supply several copies of several different multigrid styles without coordinate axes drawn in. Students can then select the appropriate style multigrid for any given problem and also select an appropriate point to insert the origin. One style of multigrid paper can be used with several different basis vector combinations. For example, the multigrid of Figure 1 can also be used for coordinates with respect to the bases $\{u_1, e_1\}$, $\{e_1, -u_2\}$, or $\{u_2, 2e_1\}$. It is even useful for the orthonormal basis $\left\{\frac{1}{\sqrt{5}}u_1, \frac{1}{\sqrt{5}}u_2\right\}$.

In closing, it is briefly noted that properties of orthonormal bases can be nicely illustrated in connection with quadratic forms. The hyperbola $0.6x^2 + 1.6xy - 0.6y^2 - 1.6\sqrt{5}x - 0.8\sqrt{5}y + 3 = 0$ can be easily plotted on the multigrid of Figure 1, and the parabola $x^2 - 2xy + y^2 - x - y + 2 = 0$ can be plotted on a multigrid based on $e_1, e_2, (1,1)^T$, and $(-1,1)^T$. The quadratic terms in these second-degree equations determine orthogonal change of basis matrices. These matrices will represent either rotations or reflections depending on whether their determinants are 1 and -1, and provide an opportunity to illustrate differences between right- and left-handed coordinate systems. Such discussions help stimulate student interest and provide links to other areas of mathematics.

Jean H. Bevis
Georgia State University
Contributed

PART 5
Matrix Forms

Introduction

Matrices can be written in many forms. Among the familiar forms are the LU factorization, the singular value decomposition and the Jordan Decomposition Theorem. There are also canonical (or echelon) forms for various types of matrices. Equivalence, similarity, orthogonal similarity, unitary similarity will all give matrix forms for certain types of matrices. A course in linear algebra would not be complete without a discussion of at least some matrix forms.

LU Factorization

The n-by-n matrix A is said to have an LU factorization if A may be written

$$A = LU$$

with L an n-by-n lower triangular and U an n-by-n upper triangular matrix. Such factorization is useful in the solution of linear systems, especially in updating schemes for iterative methods, and it can also be a convenient proof tool. The idea now frequently creeps into elementary textbooks, but seldom is there a discussion of existence. Often it is useful to require, additionally, that L or U, or both, be invertible, and all possibilities can occur. For example, $\begin{bmatrix} 0 & 1 \\ 1 & 0 \end{bmatrix}$ has no LU factorization, and $\begin{bmatrix} 0 & 0 & 0 \\ 0 & 0 & 1 \\ 1 & 0 & 0 \end{bmatrix}$ has no LU factorization with either L or U invertible.

When, then, does an LU factorization exist? It is a familiar fact that A has an LU factorization with L and U invertible if and only if the leading principal minors of A are nonzero [1]. However, if we ask only that U be invertible [2], there is another nice and much weaker condition: For each k, the vector occupying the first k entries of column $k+1$ of A must lie in the span of the columns of the leading k-by-k principal submatrix of A.

This, as with much else about LU factorization, may be seen from a partitioned point of view. Let

$$A = \begin{bmatrix} A_{11} & A_{12} \\ A_{21} & A_{21} \end{bmatrix}, \qquad \text{with } A_{11} \ k\text{-by-}k,$$

and consider the equation

$$A = \begin{bmatrix} L_{11} & 0 \\ L_{12} & L_{22} \end{bmatrix} \begin{bmatrix} U_{11} & U_{12} \\ 0 & U_{22} \end{bmatrix},$$

with L and U partitioned conformally. Inductively, assume that $k = n - 1$. Then, if U_{11} is invertible, we must have $A_{12} = L_{11}U_{12} = L_{11}U_{11}U_{11}^{-1}U_{12} = A_{11}(U_{11}^{-1}U_{12})$ in the column space of A_{11}. On the other hand, if A_{12} is in the column space of $A_{11}, A_{12} = A_{11}x$, and U_{11} is invertible (inductive hypothesis), then choose $U_{12} = U_{11}x$, $L_{12} = A_{21}U_{11}^{-1}$, and $U_{22} = 1$ (solving for L_{22}) to extend the factorization with U invertible. $\square$

There is an analogous fact when L is asked to be invertible: The vector in the first k entries of row $k+1$ of A must lie in the row space of the leading k-by-k principal submatrix of A.

The existence of an LU factorization with no invertibility requirement has long been an open question. It was settled recently by C.R. Johnson and P. Okunev.

References

1. R. Horn and C.R. Johnson, *Matrix Analysis*, Cambridge University Press, New York, 1985.
2. C. Lau and T. Markham, LU Factorizations, *Czechoslovak Math J.* **29** (1979), 546–550.

Charles R. Johnson
The College of William and Mary
Contributed

Singular Value Decomposition
The 2×2 Case

Let A be a real $n \times n$ matrix. It is a fact that there exist real orthogonal matrices V and W and a diagonal matrix Σ with nonnegative diagonal entries such that

$$A = W\Sigma V^T. \tag{1}$$

The diagonal entries of Σ are the *singular values* of A, and the factorization (1) is called the *singular value decomposition* of A.

We will prove that any 2×2 nonsingular matrix admits such a factorization.

Let $\langle \cdot, \cdot \rangle$ denote the usual inner product, and define

$$e_1(\theta) = \begin{bmatrix} \cos\theta \\ \sin\theta \end{bmatrix}, \qquad e_2(\theta) = \begin{bmatrix} -\sin\theta \\ \cos\theta \end{bmatrix}.$$

Note that $\langle e_1(\theta), e_2(\theta) \rangle = 0$ for all θ.

Let A be a given 2×2 nonsingular matrix, and let us define a function $f : \mathbf{R} \to \mathbf{R}$ by

$$f(\theta) = \langle Ae_1(\theta), Ae_2(\theta) \rangle.$$

First let us argue that there is some $\theta_0 \in [0, \pi/2]$ for which $f(\theta) = 0$. To see this, observe that $e_1(\pi/2) = e_2(0)$ and $e_2(\pi/2) = -e_1(0)$, so that $f(\pi/2) = -f(0)$. Thus if $f(0) > 0$ then $f(\pi/2) < 0$; and since f is a continuous function of θ, the intermediate value theorem implies that there is some $\theta_0 \in (0, \pi/2)$ for which $f(\theta_0) = 0$. A similar argument applies if $f(0) < 0$, and if $f(0) = 0$ then of course we simply take $\theta_0 = 0$.

The existence of θ_0 is the crux of the proof, for not only are $e_1(\theta_0)$ and $e_2(\theta_0)$ orthogonal (in fact they are orthonormal), but so are $Ae_1(\theta_0)$ and $Ae_2(\theta_0)$. The significance of this should become clear momentarily.

Let us proceed by setting $v_1 = e_1(\theta_0)$, $v_2 = e_2(\theta_0)$, $\sigma_1 = \|Av_1\|$, $\sigma_2 = \|Av_2\|$, $w_1 = Av_1/\sigma_1$ and $w_2 = Av_2/\sigma_2$. Note that w_1 and w_2 are orthonormal. Observe now that the matrices

$$V = \begin{bmatrix} v_1 & v_2 \end{bmatrix} \quad \text{and} \quad W = \begin{bmatrix} w_1 & w_2 \end{bmatrix}$$

are orthogonal.

Finally, set

$$\sigma = \begin{bmatrix} \sigma_1 & 0 \\ 0 & \sigma_2 \end{bmatrix}.$$

We then have

$$\begin{aligned} AV &= [Av_1 \quad Av_2] = [\sigma_1 w_1 \quad \sigma_2 w_2] \\ &= W\Sigma. \end{aligned} \tag{2}$$

The proof is completed by postmultiplying both sides of equation (2) by V^T to obtain

$$A = W\Sigma V^T.$$

Exercises

1. Let A be an $n \times n$ matrix, and for the sake of simplicity assume that A is non-singular. Suppose that $v_1, v_2, \ldots, v_n$ are given orthonormal vectors whose images $Av_1, Av_2, \ldots, Av_n$ are orthogonal. Show how to construct the singular value decomposition in this case. (Exercise 2 is concerned with the existence of such $v_1, \ldots, v_n$.)

2. This is for those familiar with the spectral theorem for real symmetric matrices. Prove that one may take $v_1, v_2, \ldots, v_n$ to be eigenvectors of $A^T A$ and that the singular values of A are the nonnegative square roots of the eigenvalues of $A^T A$.

Michael Lundquist
Brigham Young University
Contributed

A Simple Proof of the Jordan Decomposition Theorem for Matrices

There are several proofs of the existence of the Jordan normal form of a square complex matrix. For a discussion of these proofs and references, we refer the reader to the paper by Väliaho [1].

In this note we present a new, simple and short proof of the existence of the Jordan decomposition of an operator on a finite dimensional vector space over the complex numbers. Our proof is based on an algorithm that allows one to build the Jordan form of an operator A on an n-dimensional space if the Jordan form of A restricted to an $n-1$ dimensional invariant subspace is known.

Let A be a linear operator on a finite-dimensional vector space V over the complex numbers. Recall that a subspace of V is called *cyclic* if it is of the form

$$\text{span}\left\{\varphi, (A-\lambda)\varphi, \ldots, (A-\lambda)^{m-1}\varphi\right\}$$

with $(A-\lambda)^{m-1}\varphi \neq 0$ and $(A-\lambda)^m\varphi = 0$. Such a subspace is A-invariant and has dimension m. This follows immediately from the fact that if for some $r (r = 0, 1, \ldots, m-1)$

$$c_r(A-\lambda)^r\varphi + \cdots + c_{m-1}(A-\lambda)^{m-1}\varphi = 0 \quad \text{and} \quad c_r \neq 0,$$

then after an application of $(A-\lambda)^{m-r-1}$ to both sides of this equality we obtain

$$c_r(A-\lambda)^{m-1}\varphi = 0.$$

Idea of the proof: The argument can be reduced to two cases. In one case there is a vector g outside of an $n-1$ dimensional A-invariant subspace F of V such that $Ag = 0$. In this case $V = F \oplus \text{span}\{g\}$ and the solution is clear from the induction hypothesis on F. The difficult case is when no such g exists. It turns out that one of the cyclic subspaces of the restriction of A to F is replaced by a cyclic subspace of A in V which is larger by one dimension while keeping the other cyclic subspaces unchanged.

Observation. Suppose $W = H \oplus \text{span}\{\varphi, A\varphi, \ldots, A^{m-1}\varphi\}$ with $A^{m-1}\varphi \neq 0, A^m\varphi = 0$, where H is an A-invariant subspace of V and $A^m H = \{0\}$. Given $h \in H$, let $\varphi' = \varphi + h$. Then

$$W = H \oplus \text{span}\left\{\varphi', A\varphi', \ldots, A^{m-1}\varphi'\right\},$$

with $A^{m-1}\varphi' \neq 0$ and $A^m\varphi' = 0$. This statement follows immediately from the fact that if a linear combination of the vectors $\varphi', A\varphi', \ldots, A^{m-1}\varphi'$ belongs to H, then the same linear combination of the vectors $\varphi, A\varphi, \ldots, A^{m-1}\varphi$ also belongs to H.

Jordan Decomposition Theorem. *Let $V \neq (0)$ be a finite dimensional vector space over the complex numbers and let A be a linear operator on V. Then V can be expressed as a direct sum of cyclic subspaces.*

Proof. The proof proceeds by induction on $\dim V$. The decomposition is trivial if $\dim V = 1$. Assume that the decomposition holds for spaces of dimension $n - 1$. Let $\dim V = n$. First we assume that A is singular. Then the range $R(A)$ of A has dimension at most $n - 1$. Let F be an $n - 1$ dimensional subspace of V which contains $R(A)$. Since $AF \subset R(A) \subset F$, the induction hypothesis guarantees that F is the direct sum of cyclic subspaces

$$M_j = \mathrm{span}\left\{\varphi_j, (A - \lambda_j)\varphi_j, \ldots, (A - \lambda_j)^{m_j - 1}\varphi_j\right\}, \quad 1 \leq j \leq k.$$

The subscripts are chosen so that $\dim M_j \leq \dim M_{j+1}, 1 \leq j \leq k-1$. Define $S = \{j | \lambda_j = 0\}$. Take $g \notin F$. We claim that Ag is of the form

$$Ag = \sum_{j \in S} \alpha_j \varphi_j + Ah, \quad h \in F, \tag{10}$$

if $S \neq \emptyset$. If $S = \emptyset$, then $Ag = Ah$. To verify (1), note that $Ag \in R(A) \subset F$. Hence Ag is a linear combination of vectors of the form $(A - \lambda_j)^q \varphi_j, 0 \leq q \leq m_j - 1, 1 \leq j \leq k$. For $\lambda_j = 0$, the vectors $A\varphi_j, \ldots, A_{\varphi_j}^{m-1}$, are in $A(F)$. If $\lambda_j \neq 0$, then from $(A - \lambda_j)^{m_j}\varphi_j = 0$ and from the binomial decomposition we get that φ_j is of the form $\sum_{m=1}^{m_j} b_m A^m \varphi_j$. Thus all vectors $(A - \lambda_j)^q \varphi_j$ belong to $A(F)$ and equation (1) holds.

Let $g_1 = g - h$, where h is given in (1). Since $g \notin F$ and $h \in F, g_1 \notin F$ and from equation (1),

$$Ag_1 = \sum_{j \in S} \alpha_j \varphi_j. \tag{11}$$

If $Ag_1 = 0$, then span $\{g_1\}$ is cyclic and $V = F \oplus \mathrm{span}\{g_1\}$. Suppose $Ag_1 \neq 0$. Let p be the largest of the integers j in (2) for which $\alpha_j \neq 0$. Then for $\tilde{g} = (1/\alpha_p)g_1$,

$$A\tilde{g} = \varphi_p + \sum_{j \in S, j < p} \frac{\alpha_j}{\alpha_p}\varphi_j. \tag{12}$$

Define

$$H = \sum_{j \in S, j < p} \oplus M_j.$$

The subspace H is A-invariant and since $\dim M_j \leq \dim M_p, j < p$, it follows that $A^{m_p}(H) = \{0\}$. Thus by the observation applied to $H \oplus M_p$ and equality (3), we have

$$H \oplus M_p = H \oplus \mathrm{span}\left\{A\tilde{g}, \ldots, A^{m_p}\tilde{g}\right\}.$$

Hence,

$$F = \sum_{j \neq p} \oplus M_j \oplus \text{span} \left\{ A\tilde{g}, \ldots, A^{m_p}\tilde{g} \right\}.$$

Since $\tilde{g} \notin F$,

$$V = F \oplus \text{span} \{\tilde{g}\} = \sum_{j \neq p} \oplus M_j \oplus \text{span} \left\{ \tilde{g}, A\tilde{g}, \ldots, A^{m_p}\tilde{g} \right\}.$$

This completes the proof of the theorem under the assumption that A is singular. For the general case, let μ be an eigenvalue of A. Then $A - \mu$ is singular and by the above result applied to $A - \mu$, it follows that V is the direct sum of cyclic subspaces for A. Q.E.D.

This proof shows how to extend a Jordan form for A on an $n-1$ dimensional invariant subspace F to an n-dimensional A-invariant subspace containing F.

Note that the proof of the theorem also holds if the field of complex numbers is replaced by any algebraically closed field.

Illustrative Example. Let

$$A = \begin{pmatrix} 0 & 1 & 0 & 0 & a \\ 0 & 0 & 0 & 0 & b \\ 0 & 0 & 0 & 1 & c \\ 0 & 0 & 0 & 0 & d \\ 0 & 0 & 0 & 0 & 0 \end{pmatrix}$$

Then

$$Ae_2 = e_1, \quad Ae_1 = 0, \quad Ae_4 = e_3, \quad Ae_3 = 0.$$

We take

$$F = \text{span} \{e_1, e_2, e_3, e_4\} = \text{span} \{e_2, Ae_2\} \oplus \text{span} \{e_4, Ae_4\}.$$

Now $e_5 \notin F$ and

$$Ae_5 = ae_1 + be_2 + ce_3 + de_4 = be_2 + de_4 + A(ae_2 + ce_4).$$

If $d \neq 0$, take $\tilde{g} = e_5 - ae_2 - ce_4/d$. Then $A\tilde{g} = e_4 + (b/d)e_2$, $A^2\tilde{g} = e_3 + (b/d)e_1$ and

$$\mathbb{C}^5 = \text{span} \{e_2 Ae_2\} \oplus \text{span} \left\{ \tilde{g}, A\tilde{g}, A^2\tilde{g} \right\}.$$

If $d = 0$ and $b \neq 0$ take $\tilde{g} = e_5 - ae_2 - ce_4/b$. Then $A\tilde{g} = e_2$ and $Ae_2 = e_1$. Hence

$$\mathbb{C}^5 = \text{span} \left\{ \tilde{g}, A\tilde{g}, A^2\tilde{g} \right\} \oplus \text{span} \{e_4, Ae_4\}.$$

Finally, if $d = b = 0$ take $\tilde{g} = e_5 - ae_2 - ce_4$. then $A\tilde{g} = 0$ and

$$\mathbb{C}^5 = \text{span} \{e_2, Ae_2\} \oplus \text{span} \{e_4, Ae_4\} \oplus \text{span} \{\tilde{g}\}.$$

Reference

1. H. Väliaho, An elementary approach to the Jordan form of a matrix, *American Mathematical Monthly* **93**, No. 9 (1986), 711–714.

Israel Gohberg
Tel Aviv University

Seymour Goldberg
University of Maryland
American Mathematical Monthly **103** (1996), 157–159

Similarity of Matrices

It is well known that if A and B are $n \times n$ matrices and if they are similar over the complex numbers then they are similar over the real numbers. Although this theorem is understandable to a beginning linear algebra student, its proof is usually part of the more advanced theory of similarity invariants [2,p. 144], [1,p. 203]. The purpose of this note is to give an elementary proof that requires only some simple facts about determinants and polynomials. First, if $\det X \neq 0$, then X is nonsingular. Second, a nonzero polynomial has only finitely many roots.

Theorem. *Let A and B be real $n \times n$ matrices. If A is similar to B over the complexes, then A is similar to B over the reals.*

Proof. Suppose that $A = S^{-1}BS$, for some nonsingular complex matrix S. There are real matrices P and Q such that $S = P + iQ$. Then $(P + iQ)A = B(P + iQ)$; and since $A, B, P,$ and Q are all real, we have $PA = BP$ and $QA = BQ$. If either P or Q is nonsingular, then we are finished. Even if both P and Q are singular, we are finished if there is a real number r such that $P + rQ$ is nonsingular. For then $(P + rQ)A = B(P + rQ)$.

We know that the polynomial $p(x) = \det(P + xQ)$ is not identically zero since $p(i) = \det S \neq 0$. It follows that in any infinite set there is an element r such that $p(r) \neq 0$. In particular there is a real number r such that $p(r) \neq 0$. But then $P + rQ$ is a real nonsingular matrix and we are finished.

References

1. K. Hoffman and R. Kunze, *Linear Algebra*, Prentice Hall, Englewood Cliffs, N.J., 1961.
2. S. Perlis, *Theory of Matrices*, Addison Wesley, Reading, MA, 1952.

William Watkins
California State University, Northridge
American Mathematical Monthly **87** (1980), 300

Classifying Row-reduced Echelon Matrices

One of the fundamental theorems of linear algebra states that the row-reduced echelon form of a *given* $m \times n$ matrix is unique. Yet, there are many row-reduced echelon forms associated with the set of *all* $m \times n$ matrices, for given m and n with $n > 1$. We shall characterize them and count their number.

Recall that a matrix is said to be in *row-reduced echelon form* if the following conditions are all satisfied:

(i) In each row that does not consist entirely of zeros, the first nonzero entry is a one (known as a *leading one*).

(ii) In each column that contains a leading one of some row, all other entries are zero.

(iii) In any two rows with nonzero entries, the leading one of the higher row is farther to the left.

(iv) Any row that contains only zeros is lower than all rows that have some nonzero entries.

In any row-reduced echelon matrix, we shall refer to the (positioned) zeros required by conditions (i), (ii), and (iv) of the definition as *forced zeros*.

As an example, consider all possible 2×3 row-reduced echelon matrices. Each of these can be written as a special case of one of the following, where x and y are arbitrary.

$$\begin{bmatrix} 0 & 0 & 0 \\ 0 & 0 & 0 \end{bmatrix}, \quad \begin{bmatrix} 1 & x & y \\ 0 & 0 & 0 \end{bmatrix}, \quad \begin{bmatrix} 0 & 1 & x \\ 0 & 0 & 0 \end{bmatrix}, \quad \begin{bmatrix} 0 & 0 & 1 \\ 0 & 0 & 0 \end{bmatrix},$$

$$\begin{bmatrix} 1 & 0 & x \\ 0 & 1 & y \end{bmatrix}, \quad \begin{bmatrix} 1 & x & 0 \\ 0 & 0 & 1 \end{bmatrix}, \quad \begin{bmatrix} 0 & 1 & 0 \\ 0 & 0 & 1 \end{bmatrix}.$$

The displayed ones and zeros are *leading* ones and *forced* zeros. We shall refer to entries that are neither leading ones nor forced zeros as *undetermined* entries. Thus, we have three *types* of entries for each matrix.

Although there are infinitely many matrices represented above, there are only seven different "classes" of them. To make this idea precise, we shall say that two row-reduced echelon matrices are *type-equivalent* if each pair of corresponding entries is of the same type. For example, the matrices

$$\begin{bmatrix} 1 & 0 & 2 & 3 \\ 0 & 1 & 4 & 5 \\ 0 & 0 & 0 & 0 \end{bmatrix} \quad \text{and} \quad \begin{bmatrix} 1 & 0 & 1 & 0 \\ 0 & 1 & 8 & 1 \\ 0 & 0 & 0 & 0 \end{bmatrix}$$

are type-equivalent, but the matrices

$$\begin{bmatrix} 1 & 0 & 0 \\ 0 & 1 & 0 \end{bmatrix} \quad \text{and} \quad \begin{bmatrix} 1 & 0 & 0 \\ 0 & 0 & 1 \end{bmatrix}$$

are not. In the latter case, the entries in the following positions are of different types: $(1,2),(1,3),(2,2)$ and $(2,3)$.

The relation induced by "type-equivalence" is an equivalence relation—it is symmetric, reflexive and transitive. It therefore partitions the set of all $m \times n$ row-reduced echelon matrices into equivalence classes. The above pair of 3×4 matrices belongs to the same equivalence class, whereas the above pair of 2×3 matrices does not. In this context, the seven 2×3 row-reduced echelon matrices displayed earlier are the seven equivalence classes that partition the set of all 2×3 row-reduced echelon matrices. Note that

$$7 = \binom{3}{0} + \binom{3}{1} + \binom{3}{2}.$$

In general, the number of distinct type-equivalent $m \times n$ matrices (that is, the number of equivalence classes induced by the "type" relation) is given by

$$N(n,m) = \sum_{k=0}^{\min(m,n)} \binom{n}{k}.$$

To verify this, first observe that the positions of the leading ones in any row-reduced echelon matrix determine the type of *all* the entries in that matrix. (They determine the positions of the forced zeros, and hence those of the undetermined entries.) Thus, all we need show is that the number of ways that the leading ones can be arranged is given by the formula above. Our approach will be to do this for matrices of *rank k* (that is, with k nonzero rows in their reduced form) and then sum the results from rank 0 to rank $\min(m,n)$ (the largest possible).

Suppose A is an $m \times n$ row-reduced echelon matrix of rank k. Then A has exactly k leading ones. These leading ones, located in the first k rows of A, must occur in k distinct columns. Once the columns are specified, the positions of the leading ones are completely determined since they form "stair steps" down to the right. Since there are $\binom{n}{k}$ ways of choosing k objects from a collection of n distinct objects, there are $\binom{n}{k}$ ways of positioning the leading ones. Thus, there are exactly $\binom{n}{k}$ equivalence classes for $m \times n$ row-reduced echelon matrices of rank k. Summing over $k = 0, 1, \dots, \min(m,n)$ completes the proof.

As an example, observe that the number of distinct type-equivalent 3×3 row-reduced echelon matrices is

$$N(3,3) = \binom{3}{0} + \binom{3}{1} + \binom{3}{2} + \binom{3}{3} = 8.$$

It is no coincidence that the answer turned out to be 2^3. Indeed, the number of distinct type-equivalent square matrices of order n is equal to 2^n. This follows immediately from

$$N(n,n) = \sum_{k=0}^{n} \binom{n}{k},$$

since $(1+x)^n = \sum_{k=0}^{n} \binom{n}{k} x^k$ yields $2^n = \sum_{k=0}^{n} \binom{n}{k}$.

Stewart Venit and Wayne Bishop
California State University, Los Angeles
College Mathematics Journal **17** (1986), 169–170

PART 6
Polynomials and Matrices

Introduction

One of the early connections between polynomials and matrices is the well known Cayley-Hamilton Theorem. Polynomial equations in which the coefficients are matrices (so they are called polynomial matrix equations) have been studied extensively in the past sixty or more years. Many results exist both over the real and complex field as well as over finite fields. Matrix methods have been applied to the row (and column) reduction of systems of linear equations and to finding the g.c.d. of sets of polynomials. Indeed, the connections between polynomials and matrices are many and varied.

On the Cayley-Hamilton Theorem

Let A be an n-by-n matrix over an arbitrary field and let $p(x) = \det(xI - A)$ denote the characteristic polynomial of A. Then

$$p(A) = 0.$$

Proof: Define $F(x) = \mathrm{adj}(xI - A)$, in which adj is the classical adjoint (or matrix of cofactors). Then, $F(x)$ is an n-by-n matrix, whose entries are polynomials in x of degree at most $n - 1$. We have

$$F(x)(xI - A) = p(x)I.$$

It follows that

(a) every entry of $F(x)(xI - A)$ is a polynomial divisible by $p(x)$; and

(b) over the field of rational functions in x, $F(x)$ is invertible, as $F(x)^{-1} = \dfrac{1}{p(x)}(xI - A)$.

From (a) each entry of $xF(x) - F(x)A$ is divisible by $p(x)$. Thus, each entry of $xF(x)A - F(x)A^2$, or equivalently, $x^2 F(x) - F(x)A^2$ is divisible by $p(x)$. Continuing in the same way, all entries of

$$x^k F(x) - F(x)A^k$$

are divisible by $p(x)$ for all positive integers k. Multiplication by the kth coefficient of p and summing on k yields that

$$p(x)F(x) - F(x)p(A)$$

has all entries divisible by $p(x)$. Thus all entries of $F(x)p(A)$ are divisible by $p(x)$. However, all entries of $F(x)p(A)$ are of degree at most $n - 1$ and, as $p(x)$ has degree n, must be 0. Finally, since by (b) $F(x)$ is invertible, $p(A)$ is 0. $\square$

Robert Reams
University College, Galway (Ireland)
Contributed

On Polynomial Matrix Equations

In his stimulating article "What Do I Know? A Study of Mathematical Self-Awareness" [*CMJ* **16** (January 1985), 22-41], Phil Davis asked on page 26 if the equation

$$X^3 + X^2 + \begin{bmatrix} 0 & 1 \\ 0 & 0 \end{bmatrix} X = \begin{bmatrix} 1 & 2 \\ 3 & 4 \end{bmatrix} \tag{1}$$

has a real 2×2 matrix solution. I wrote Phil that of course it does, and (tongue-in-cheek) pulled the rabbit out of the hat:

Let λ and μ be distinct real solutions of

$$(t^3 + t^2 - 1)(t^3 + t^2 - 4) + 3(t - 2) = 0, \tag{2}$$

and set

$$P = \begin{bmatrix} \lambda^3 + \lambda^2 - 4 & \mu^3 + \mu^2 - 4 \\ 3 & 3 \end{bmatrix}. \tag{3}$$

Behold, a solution of (1):

$$X = P \begin{bmatrix} \lambda & 0 \\ 0 & \mu \end{bmatrix} P^{-1}. \tag{4}$$

This certainly seems like magic, and its explanation should provide some material for enrichment projects and student research. (For a first exercise, prove that (2) has exactly two real solutions.) Let us start by examining the class of problems of which (1) is an example.

Let $A_0, A_1, \ldots, A_r$ (where $A_r = I$) be fixed $n \times n$ complex matrices and consider the equation

$$\sum_{j=0}^{r} A_j X^j = 0 \tag{5}$$

for an $n \times n$ unknown matrix X. (Note that because matrices do not usually commute, (5) is not the most general polynomial matrix equation; all powers of X are on the right. Ask your students to experiment with

$$X^3 + AX^2 + B = 0, \qquad X^3 + XAX + B = 0, \qquad X^3 + X^2 A + B = 0$$

147

for instance.)

To analyze equation (5), suppose we have a solution X. Since the complex square matrix X has characteristic vectors, assume that $X\mathbf{v} = \lambda\mathbf{v}$ for a nonzero column vector $\mathbf{v}$ and complex number λ. Then $X^j\mathbf{v} = \lambda^j\mathbf{v}$ for $j = 0, 1, \ldots, r$. Hence, by (5),

$$\left(\sum_{j=0}^{r} A_j \lambda^j\right) \mathbf{v} = \mathbf{0}. \tag{6}$$

Since $\mathbf{v} \neq \mathbf{0}$, it follows λ is a solution of

$$\det\left(\sum_{j=0}^{r} A_j t^j\right) = 0. \tag{7}$$

Because $A_r = I$, the determinant is a polynomial of degree nr in the (scalar) variable t.

Now let us try to piece together a matrix solution of (5). Suppose that $\lambda_1, \lambda_2, \ldots, \lambda_n$ are n, not necessarily distinct, solutions of (7) and that $\mathbf{v}_1, \ldots, \mathbf{v}_n$ are nonzero column vectors such that

$$\left(\sum_{j=0}^{r} A_j \lambda_i^j\right) \mathbf{v}_i = \mathbf{0} \tag{8}$$

for $i = 1, 2, \ldots, n$. Such nonzero vectors exist because by (7), each matrix

$$\sum_{j=0}^{r} A_j \lambda_i^j$$

is singular. Next we define $n \times n$ matrices

$$P = [\mathbf{v}_1, \ldots, \mathbf{v}_n] \quad \text{and} \quad D = \begin{bmatrix} \lambda_1 & & 0 \\ & \ddots & \\ 0 & & \lambda_n \end{bmatrix}, \tag{9}$$

and compute

$$PD^j = [\mathbf{v}_1, \ldots, \mathbf{v}_n] \begin{bmatrix} \lambda_i^j & & 0 \\ & \ddots & \\ 0 & & \lambda_n^j \end{bmatrix} = \left[\lambda_1^j \mathbf{v}_1, \ldots, \lambda_n^j \mathbf{v}_n\right]. \tag{10}$$

It follows that

$$\sum_{j=0}^{r} A_j PD^j = \sum_{j=0}^{r} A_j \left[\lambda_1^j \mathbf{v}_1, \ldots, \lambda_n^j \mathbf{v}_n\right]$$

$$= \left[\sum_{j=0}^{r} A_j \lambda_1^j \mathbf{v}_1, \ldots, \sum_{j=0}^{r} A_j \lambda_n^j \mathbf{v}_n\right].$$

By (8), each column in this $n \times n$ matrix equals $\mathbf{0}$. Thus the n vector equations in (8) are equivalent to the single neat matrix equation

$$\sum_{j=0}^{r} A_j P D^j = 0 \tag{11}$$

So far, the n vectors $\mathbf{v}_i$ are nonzero; that is all. But suppose that somehow we can choose λ's and $\mathbf{v}$'s satisfying (7) and (8), and such that $\mathbf{v}_1, \ldots, \mathbf{v}_n$ are also *linearly independent* (a big "suppose"). Then the matrix P in (9) is nonsingular; so we can multiply (11) by P^{-1} to obtain

$$\sum_{j=0}^{r} A_j (P D^j P^{-1}) = 0. \tag{12}$$

But $P D^j P^{-1} = (P D P^{-1})^j$. Hence if we set

$$X = P D P^{-1}, \tag{13}$$

then

$$\sum_{j=0}^{r} A_j X^j = 0. \tag{14}$$

Thus X is a solution of (5).

To summarize: We wish to solve (5) for an $n \times n$ matrix X. We find n complex solutions $\lambda_1, \ldots, \lambda_n$ of (7), and corresponding nonzero column vector solutions $\mathbf{v}_1, \ldots, \mathbf{v}_n$ of (8). If we can do this in such a way that $\mathbf{v}_1, \ldots, \mathbf{v}_n$ are linearly independent, then the matrix X in (13), where P and D are defined by (9), is a solution of (5). This discussion raises a lot of questions, but first let us look at some examples.

Example 1. Consider Phil Davis's equation (1) for 2×2 matrices. Equation (7) of our analysis specializes to equation (2); so (2) really does come from analysis, not from a hat! It turns out that (2) has precisely two real solutions (and they are simple):

$$\lambda_1 \approx -1.4834075 \quad \text{and} \quad \lambda_2 \approx 1.3794399.$$

The corresponding equation (8) for $\mathbf{v}_1$ is

$$\begin{bmatrix} -2.0637371 & -3.4834075 \\ -3.0000000 & -5.0637371 \end{bmatrix} \mathbf{v}_1 \approx \mathbf{0}.$$

(The 2×2 matrix is indeed very close to singular.) Since $\mathbf{v}_1$ is only determined up to a factor, we may choose

$$\mathbf{v}_1 \approx \begin{bmatrix} -5.0637371 \\ 3.0000000 \end{bmatrix}.$$

This is surely accurate to 6 places. Similarly, we may choose

$$\mathbf{v}_2 \approx \begin{bmatrix} 0.5277290 \\ 3.0000000 \end{bmatrix}.$$

Since $\mathbf{v}_1$ and $\mathbf{v}_2$ are linearly independent, $P = [\mathbf{v}_1, \mathbf{v}_2]$ is nonsingular. It is now an interesting (computer) exercise to calculate the $X = PDP^{-1}$ in (13) and check that it solves (1) to reasonable accuracy.

Example 2. Equation $X^2 - I = 0$ for 2×2 matrices. Equation (7) specializes to

$$(t^2 - 1)^2 = 0,$$

which has $\lambda_1 = -1$ and $\lambda_2 = 1$ as its only distinct solutions. In both cases $\lambda^2 I - I = 0$, so $\mathbf{v}_1$ and $\mathbf{v}_2$ are arbitrary nonzero vectors. We have lots of linearly independent choices, suggesting many solutions of $X^2 = I$. How many in addition to I and $-I$?

Example 3. Equation $X^2 = 0$ for 2×2 matrices. Equation (7) becomes $t^4 = 0$, with unique solution $\lambda = 0$. The only solution $X = PDP^{-1}$ this method produces is $X = 0$, because $D = 0$. But there are lots of other 2×2 matrices X such that $X^2 = 0$:

$$\begin{bmatrix} 0 & a \\ 0 & 0 \end{bmatrix}, \begin{bmatrix} 0 & 0 \\ b & 0 \end{bmatrix}, P \begin{bmatrix} 0 & 1 \\ 0 & 0 \end{bmatrix} P^{-1}.$$

Thus our method may fail to produce all solutions.

Example 4. For 3×3 matrices, consider the equation

$$X^2 = \begin{bmatrix} 0 & 0 & 1 \\ 0 & 0 & 0 \\ 0 & 0 & 0 \end{bmatrix}.$$

Equation (7) becomes $t^6 = 0$, whose only solution is $\lambda = 0$. Thus the method fails to produce *any* solution. But

$$X = \begin{bmatrix} 0 & 1 & 0 \\ 0 & 0 & 1 \\ 0 & 0 & 0 \end{bmatrix}$$

is a solution. Why does the method fail?

Example 5. Consider the equation

$$X^2 = \begin{bmatrix} 0 & 1 \\ 0 & 0 \end{bmatrix}$$

for 2×2 matrices. Again, our method fails to produce any solution. However, in this case, there is none! Why? Do the ideas of this article help to prove this?

Further questions. The preceding problem can be used as a basis for student investigations that go beyond the usual textbook exercises. We offer a few questions worth exploring.

(i) Suppose equation (7) has nr distinct roots. Then does our method produce at least one solution to (5)? All solutions?

(ii) Can we find conditions on equation (5) that guarantee failure of our method: no linearly independent solutions of (8)?

(iii) Explore some examples in which (5) factors into commuting factors—as, for instance, $(X - I)(X - A) = 0$.

For other material on the subject of matrix equations, the reader may consult the author's article "Analytic Solutions of Matrix Equations" [Linear and Multilinear Algebra 2 (1974) 241–243] and the treatise *Matrix Polynomials* by I. Gohberg, P. Lancaster, and L. Rodman [Academic Press, 1982]; relevant material is found around pages 113 and 125.

Harley Flanders
Jacksonville University
College Mathematics Journal **17** (1986), 388–391

The Matrix Equation $X^2 = A$

The purpose of this note is to give a useful procedure for solving the matrix equation $X^2 = A$.

In response to the curiosity of sophomores in linear algebra courses I have accumulated a number of exercises and theorems involving polynomial equations in matrices suitable for an elementary course. Chapter 7 of a book [5] of C.C. MacDuffee provides a summary of basic results and is a good source of references for matrix equations up to 1933. It is adequate for enriching an elementary course. Matrix equations is still an active research field (cf. [2],[4], for example) but I am unaware of a recent survey. According to [6, Chapter 2], the solution of the matrix equation $\sum_{i=0}^{n} A_i X^i = 0$ is equivalent to the search for the right divisors $I\lambda - X$ of $\sum_{i=0}^{n} A_i \lambda^i$. In this connection the paper [3] is of interest.

Returning to the simple quadratic $X^2 = A$, a classical theorem [1, p. 299] ensures a (nonsingular) solution to the equation if A is nonsingular. I now describe a method, once suggested to me by W.E. Roth, that reduces the problem of solving $X^2 = A$ to linear equations regardless of the rank of A.

If X is a solution of the equation (examples for which there is no solution are easily devised), then

$$X^2 - \lambda^2 I = A - \lambda^2 I;$$

hence

$$(X - \lambda I)(X + \lambda I) = A - \lambda^2 I.$$

If $\phi(\lambda)$ denotes the characteristic function of a solution, then

$$\phi(\lambda)\phi(-\lambda) = \det(A - \lambda^2 I) \qquad (*)$$

and so the characteristic function of a solution, if one exists, must be a divisor of $\det(A - \lambda^2 I)$. For each of the possible solutions $\phi(\lambda)$ of (*) one secures the equation $\phi(X) = 0$ in which all even powers of X may be replaced by powers of the known matrix A to reduce the equation to a linear matrix equation.

As an example, consider the equation

$$X^2 = \begin{bmatrix} 9 & 0 & -8 \\ 5 & 4 & -5 \\ 0 & 0 & 0 \end{bmatrix}$$

For this A,

$$\phi(\lambda)\phi(-\lambda) = \det(A - \lambda^2 I) = -\lambda^2(\lambda^2 - 9)(\lambda^2 - 4)$$

and the possible characteristic functions of solutions are $\phi_1(\lambda) = \lambda(\lambda - 3)(\lambda - 2)$, $\phi_1(-\lambda)$, $\phi_2(\lambda) = \lambda(\lambda + 3)(\lambda - 2)$, $\phi_2(-\lambda)$, $\phi_3(\lambda) = \lambda(\lambda + 3)(\lambda + 2)$, $\phi_3(-\lambda)$, $\phi_4(\lambda) = \lambda(\lambda - 3)(\lambda + 2)$, $\phi_4(-\lambda)$. In this problem the solutions given by the last four functions are the solutions given by the first four functions.

Consider

$$\phi_1(X_1) = X_1^3 - 5X_1 + 6X_1 = AX_1 - 5A + 6X_1 = (A + 6I)X_1 - 5A = 0.$$

This linear equation has the unique solution

$$X_1 = \begin{bmatrix} 3 & 0 & -8/3 \\ 1 & 2 & -7/6 \\ 0 & 0 & 0 \end{bmatrix},$$

which satisfies the given equation. Similarly, $\phi_2(X_2) = 0$ gives

$$X_2 = \begin{bmatrix} -3 & 0 & 8/3 \\ -5 & 2 & 25/3 \\ 0 & 0 & 0 \end{bmatrix}.$$

These two matrices and their negatives are the four solutions of the given quadratic equation.

References

1. Maxime Bôcher, *Introduction to Higher Algebra*, Macmillan, New York, 1907.
2. Harley Flanders, and Harald K. Wimmer, On the matrix equation $AX - XB = C$ and $AX - YB = C$, *SIAM J. Appl. Math.* **32** (1977) 707–710.
3. I. Gohberg, P. Lancaster, and L. Rodman, Spectral analysis of matrix polynomials, I, *Linear Alg. and Appl.* **20** (1978), 1–44.
4. H. Langer, Factorization of operator pencils, *Acta Sci. Math. Szeged* **38** (1976) 83–96.
5. C.C. MacDuffee, *The Theory of Matrices*, Chelsea, New York, 1946.
6. N.H. McCoy, *Rings and Ideals*, MAA Carus Monographs, Baltimore, 1948.

W.R. Utz
University of Missouri, Columbia
American Mathematical Monthly **86** (1979), 855–856

Editors' Note: Exact characterization of which (complex) matrices have (complex) square roots is known. It hinges upon the Jordan structure of the eigenvalue 0 and is discussed in a few standard references (e.g., R. Horn and C.R. Johnson, *Topics in Matrix Analysis*, Cambridge University Press, New York, 1991, p. 469 and following). An early reference is G. Cross and P. Lancaster, Square Roots of Complex Matrices, *Linear and Multilinear Algebra* **1** (1974), 289–293.

Where Did the Variables Go?

In the study of a branch of mathematical control theory called Liapunov stability theory, the following matrix equation arises:

$$A^T P + PA = -Q. \tag{13}$$

In equation (1), A is a real $n \times n$ matrix, Q is an arbitrary real symmetric positive definite $n \times n$ matrix, and P is a real symmetric $n \times n$ matrix to be determined. The unknown elements p_{ij} of P are those on and above the principal diagonal, the remaining elements then being obtained from the symmetric nature of P. There are therefore n unknowns $p_{11}, p_{22}, \ldots, p_{nn}$ on the principal diagonal, $n - 1$ unknowns $p_{12}, p_{23}, \ldots, p_{n-1,n}$ on the line parallel to this, and so on, ending up with the single element p_{1n}. The total number of unknowns is therefore

$$n + (n - 1) + \cdots + 2 + 1 = \frac{1}{2}n(n + 1).$$

By equating elements in the same upper triangle (on and above the principal diagonal) on both sides of (1) we produce exactly $\frac{1}{2}n(n+1)$ linear equations to be solved for the $\frac{1}{2}n(n+1)$ unknown elements. These equations will have a unique solution if and only if there are no eigenvalues λ_i, λ_j of A such that $\lambda_i + \lambda_j = 0$; this will certainly be satisfied if A is nonsingular, so that all $\lambda_i \neq 0$. In applications it is of interest to determine when the λ_i all have negative real parts, and this is the case if and only if P is positive definite. For further details, including the situation when A has complex elements, see [1].

What is of interest here, however, is the somewhat surprising fact that P can be obtained by solving only $\frac{1}{2}n(n - 1)$ linear equations for $\frac{1}{2}n(n - 1)$ unknowns, a reduction of n in the number of variables—hence the title of this note. To see how this comes about, notice that $(PA)^T = A^T P^T = A^T P$ since P is symmetric, so we can write (1) as

$$(PA)^T + PA = -Q,$$

or equivalently

$$\left(PA + \frac{1}{2}Q\right)^T + PA + \frac{1}{2}Q = 0, \tag{14}$$

where we have used $Q = Q^T$. It follows from (2) that

$$PA + \frac{1}{2}Q = S, \tag{15}$$

where S is a real skew symmetric matrix, which by definition satisfies $S^T = -S$. Assuming that A is nonsingular, we can rearrange equation (3) to give

$$P = \left(S - \frac{1}{2}Q\right) A^{-1}. \tag{16}$$

Next, multiply (3) on the left by A^T to obtain

$$A^T P A + \frac{1}{2} A^T Q = A^T S, \tag{17}$$

and transpose both sides of this equation, producing

$$A^T P A + \frac{1}{2} Q A = -S A, \tag{18}$$

where we have used the results

$$\left(A^T Q\right)^T = Q^T \left(A^T\right)^T = QA,$$
$$\left(A^T S\right)^T = S^T \left(A^T\right)^T = -SA.$$

Finally, subtracting (6) from (5) gives

$$A^T S + S A = \frac{1}{2} \left(A^T Q - QA\right). \tag{19}$$

Since the principal diagonal of a real skew symmetric matrix is identically zero, the upper triangle of S contains n fewer unknown elements than for a symmetric matrix; that is

$$\frac{1}{2} n(n+1) - n = \frac{1}{2} n(n-1)$$

in total. Equation (7) therefore represents $\frac{1}{2} n(n-1)$ linear equations for the elements of S. By the assumption that A is nonsingular these equations have a unique solution, and P is then given by (4).

References

1. S. Barnett, *Polynomials and Linear Control Systems*, Marcel Dekker, New York, 1983.

2. S. Barnett and C. Storey, Stability analysis of constant linear systems by Lyapunov's second method, *Electronics Letters* **2** (1966), 165-166.

3. S. Barnett, Simplification of the Liapunov matrix equations $A^T P A - P = -Q$, *IEEE Trans. Aut. Control* **AC-19** (1974), 446–447.

4. S. Barnett, Simplification of certain linear matrix equations, *IEEE Trans. Aut. Control* **AC-21** (1976), 115–116.

Stephen Barnett
University of Essex
Contributed

A Zero-Row Reduction Algorithm for Obtaining the gcd of Polynomials

The Euclidean algorithm is traditionally employed to find the gcd (greatest common divisor) of two polynomials. Consider two polynomials, $f(x)$ and $g(x)$, where $\deg f \geq \deg g$. The Euclidean algorithm takes us through a sequence of polynomial divisions to obtain

$$f = gq_1 + r_1,$$
$$g = r_1 q_2 + r_2,$$
$$r_1 = r_2 q_3 + r_3,$$
$$\cdots,$$
$$r_{m-1} = r_m q_{m+1},$$

where $\deg g > \deg r_1 > \cdots > \deg r_m$.

The last division, which produces a zero remainder, yields r_m as the gcd of f and g. Since $\gcd(f_1, f_2, f_3) = \gcd(\gcd(f_1, f_2), f_3)$, this procedure can be extended to find the gcd of a large number of polynomials. However, using the Euclidean algorithm in the fashion illustrated above for a large set of polynomials of high degree would be tedious. In this note, we present a zero-row reduction algorithm that is based on properties of the greatest common divisor. These properties have been recognized and used in computing gcds of integers and polynomials by Brown [2], Chrystal [3], and Knuth [4], among others.

Properties of gcds. Consider two polynomials

$$f(x) = a_0 + a_1 x + a_2 x^2 + \cdots + a_n x^n$$

and

$$g(x) = b_0 + b_1 x + b_2 x^2 + \cdots + b_t x^t$$

(1)

with real coefficients and with $a_n \neq 0, b_t \neq 0, t \leq n$. If $d(x)$ is the gcd of $f(x)$ and $g(x)$, then the following are true:

1. $d(x)$ is the gcd of $k_1 f(x)$ and $k_2 g(x)$ for nonzero real k_1, k_2.

2. $d(x)$ is the gcd of $f(x) + kg(x)$ and $g(x)$ for real k.

3. If $a_0 = 0$ and $b_0 \neq 0$, then $d(x)$ is the gcd of $g(x)$ and the polynomial $f(x)/x = a_1 + a_2 x + a_3 x^2 + \cdots + a_n x^{n-1} + 0x^n$.

157

4. If $a_0 = b_0 = 0$, then $d(x)$ contains a factor x and $d(x)/x$ is the gcd of $f(x)/x$ and $g(x)/x$.

Zero-row reduction algorithm. Given a pair of polynomials $f(x)$ and $g(x)$, we apply the above properties to obtain another pair of polynomials having the same gcd as the original pair. A systematic application of these properties, together with careful selection of the nonzero factors k, k_1, and k_2, results in a sequence of polynomial pairs having decreasing degrees (see property 3) and leading ultimately to a pair $d(x)$ and 0. For computational convenience, we will represent the pair of polynomials in (1) by the following 2 by $n + 1$ matrix C:

$$C = \begin{bmatrix} a_0 & a_1 & \cdots & a_n \\ b_0 & b_1 & \cdots & b_n \end{bmatrix}. \tag{2}$$

We begin the gcd algorithm by removing any leading columns of zeros from the matrix C to obtain a matrix C' whose first column is nonzero. If the first K columns of C are zero, but the $(K + 1)$st column is nonzero, then x^K is a factor of $\gcd(f(x), g(x))$, and we can compute the rest of the gcd by examining C' (property 4).

Now let us define the *normalization* of a row to be the operation that shifts a row to the left as many positions as necessary in order to eliminate all leading zeros and then divides the entire row by its leading (nonzero) coefficient. For example, the normalization of (0,0,2,6,0,4,18,0) is (1,3,0,2,9,0,0,0). By properties 1 and 3, we can normalize each row in C' and the gcd of the polynomials represented by the rows of the new matrix will be the same as $\gcd(f(x), g(x))/x^K$.

Property 2 can now be used to change the first entry in row 2 to zero by replacing row 2 with row 2 minus row 1. We now return to the normalization process and continue to reduce the number of nonzero entries in the rows until one of the two rows consists entirely of zeros. At that point, $\gcd(f(x), g(x))/x^K$ is represented by the remaining nonzero row of the matrix.

Pseudocode for Gcd2, the algorithm for finding the gcd of two polynomials, is presented below:

Algorithm Gcd2(C)

C is the 2 by $n + 1$ coefficient matrix of two given polynomials.

BEGIN

    p=number of leading zeros in row 1 of C
    q=number of leading zeros in row 2 of C
    K=minimum (p, q)

                                     $(x^K$ is a factor of the gcd$)$

WHILE (both rows of C are nonzero)

Normalize both rows (shift each row left to remove all leading zeros and divide each coefficient by the leading nonzero element).

Let row A be the row representing the polynomial with the smaller degree (the row that has the maximum number of trailing zeros) and row B, the other polynomial. Replace row B with row $B-$row A (elementwise subtraction).

ENDWHILE

gcd is x^K times the polynomial represented by the remaining nonzero row (shift the nonzero row K places to the right).

END Gcd2.

Using Gcd2 and the fact that $\gcd(f_1, f_2, f_3) = \gcd(\gcd(f_1, f_2), f_3)$, we could devise an iterative algorithm to find the gcd of a set of m polynomials, $f_1, f_2, \ldots, f_m$. However, a more efficient algorithm can be devised, similar to the above algorithm Gcd2. Instead of an iterative and sequential approach involving two polynomials at a time, we work on the entire set of m polynomials and a matrix with m rows using an approach that can be characterized as *parallel* (and, in fact implementable as such). The following annotated example serves to explain the algorithm.

Example. Find the gcd of the polynomials

$$
\begin{aligned}
& x^2 + 3x^3 + 3x^4 + x^5 \\
& x + 2x^2 + x^3 \\
& x \quad\quad - x^3 \\
& x + x^2.
\end{aligned}
$$

1. Coefficient matrix

```
0  0  1   3  3  1
0  1  2   1  0  0
0  1  0  -1  0  0
0  1  1   0  0  0
```

2. Determine $K : K = 1$

```
0  0  1   3  3  1
0  1  2   1  0  0
0  1  0  -1  0  0
0  1  1   0  0  0
```

3. Normalize rows

```
1  3   3  1  0  0
1  2   1  0  0  0
1  0  -1  0  0  0
1  1   0  0  0  0
```

4. Switch lowest degree polynomial with top row

```
1  1   0  0  0  0
1  2   1  0  0  0
1  0  -1  0  0  0
1  3   3  1  0  0
```

5. Perform row reduction and remove zero rows, if any

$$
\begin{array}{cccccc}
1 & 1 & 0 & 0 & 0 & 0 \\
0 & 1 & 1 & 0 & 0 & 0 \\
0 & -1 & -1 & 0 & 0 & 0 \\
0 & 2 & 3 & 1 & 0 & 0
\end{array}
$$

6. Normalize rows, switch lowest degree polynomial with top row

$$
\begin{array}{cccccc}
1 & 1 & 0 & 0 & 0 & 0 \\
1 & 1 & 0 & 0 & 0 & 0 \\
1 & 1 & 0 & 0 & 0 & 0 \\
1 & 1.5 & 0.5 & 0 & 0 & 0
\end{array}
$$

7. Perform row reduction

$$
\begin{array}{cccccc}
1 & 1 & 0 & 0 & 0 & 0 \\
0 & 0 & 0 & 0 & 0 & 0 \\
0 & 0 & 0 & 0 & 0 & 0 \\
0 & 0.5 & 0.5 & 0 & 0 & 0
\end{array}
$$

8. Remove zero rows, if any

$$
\begin{array}{cccccc}
1 & 1 & 0 & 0 & 0 & 0 \\
0 & 0.5 & 0.5 & 0 & 0 & 0
\end{array}
$$

9. Normalize rows, switch lowest degree polynomial with top row

$$
\begin{array}{cccccc}
1 & 1 & 0 & 0 & 0 & 0 \\
1 & 1 & 0 & 0 & 0 & 0
\end{array}
$$

10. Perform row reduction

$$
\begin{array}{cccccc}
1 & 1 & 0 & 0 & 0 & 0 \\
0 & 0 & 0 & 0 & 0 & 0
\end{array}
$$

11. Remove zero rows

$$
\begin{array}{cccccc}
1 & 1 & 0 & 0 & 0 & 0
\end{array}
$$

The gcd is $x + x^2 = x(1 + x)$.

The algorithm is easy to implement on a computer. The details depend on the programming language chosen. We used APL2 [1] because it supports nested arrays of mixed data types and powerful operators to manipulate these arrays. Operations such as "remove all zero rows from an array" can be executed with one or two lines of APL2 code without looping. A certain degree of parallelism is implicit in the APL2 language which encourages the programmer to design and implement his/her algorithm in a way different from what might result by using an inherently sequential language. At the machine level, multiprocessor systems can be programmed to automatically allocate all or some of the available processors to perform the parallel tasks.

References

1. J.A. Brown, S. Pakin, and R.P. Polivka, *APL2 at a Glance*, Prentice Hall, NJ, 1988.

2. W.S. Brown, Euclid's algorithm and the computation of polynomial greatest common divisors, *Journal of the Association for Computing Machinery* **18** (1971), 478–504.

3. G. Chrystal, *Algebra, An Elementary Text Book*, Chelsea, NY, 1886.

4. D.E. Knuth, *The Art of Computer Programing: Seminumerical Algorithms* (Volume 2), Addison Wesley, MA, 1969.

Sidney H. Kung
Jackson University

Yap S. Chua
University of North Florida
College Mathematics Journal **21** (1990), 138–141

PART 7
Linear Systems, Inverses and Rank

Introduction

The single most useful notion of linear algebra is that of a linear system. And, with linear systems, we often associate the idea of rank of a matrix and the variety of ideas of inverses and matrices.

The articles in this chapter study these concepts in their purest, simplest form—the matrix equation $Ax = b$—and in a wide variety of significantly different variants, from morphisms of categories to systems with interval data, to other numerical aspects of the solution of linear systems. We note one article in this section discusses the fact that (in friendly settings) Row and Column Rank Are Always Equal and the article by A.J. Berrick and M.E. Keating, Rectangular Invertible Matrices, *American Mathematical Monthly*, **104** (1997), 297–302, which discusses among other things the fact that (in less friendly settings) there can exist rectangular (nonsquare) matrices X and Y for which $XY = I$ and $YX = I$. This interesting article is of a more advanced nature than others we have selected to appear.

Left and Right Inverses

Theory for the existence and calculation of the right inverse of an $n \times n$ matrix A is well developed in most elementary textbooks via row operations and reduced echelon form. Uniqueness is often also explained. Though frequently mentioned, the fact that a right inverse is also a left inverse is not often explained. Several of our colleagues (including Jim Wahab and Alan Tucker) reminded us that this may also be done nicely, though with some subtlety, via the technology of row reduction and the echelon form. The statement that B is a right inverse of A $(AB = I$ or B is a solution to $AX = I)$ means that the augmented matrix

$$[A \quad I] \text{ row reduces to } [I \quad B]. \tag{1}$$

Equivalently, there is a product E of elementary matrices such that

$$E[A \quad I] = [I \quad B] \tag{2}$$

or

$$[EA \quad E] = [I \quad B]. \tag{2'}$$

However, because of the reversibility of row operations and the unimportance of the arrangement of columns, (1) implies that

$$[B \quad I] \text{ row reduces to } [I \quad A], \tag{3}$$

which in turn means that $BA = I$ or B is a *left* inverse of A. This may also be seen from $(2')$ as $E = B$ and $EA = I$.

The Editors
Contributed

167

Row and Column Ranks Are Always Equal

The proofs I have seen in elementary linear algebra books that the row and column ranks of an $m \times n$ matrix are always equal either depend on understanding the linear transformation represented by the matrix, or involve consideration of the corresponding system of equations with an attendant confusing array of subscripts, or both. The following proof is based only on the definition of the ranks as dimensions of the row and column spaces. It makes use of the two "important facts about matrix multiplication" given in the first section of this volume—which facts, I agree, are extremely useful in the first linear algebra course. The proof is based on a rank decomposition of A.

Theorem: *If A is an $m \times n$ matrix, then the row rank of A is equal to the column rank of A.*

Proof: Let the row rank of A be r, and the column rank c. Then there is a set of c columns of A that are linearly independent and that span the column space of A. Let B be an $m \times c$ matrix with these columns. For $j = 1, 2, \ldots, n$, the *jth* column of A may be written

$$Col_j(A) = d_{1j}Col_1(B) + d_{2j}Col_2(B) + \cdots + d_{cj}Col_c(B) = Col_j(BD),$$

where $D = [d_{ij}]$ is $c \times n$. Hence $A = BD$, and so for $i = 1, 2, \ldots, m$,

$$Row_i(A) = Row_i(BD) = b_{i1}Row_1(D) + b_{i2}Row_2(D) + \cdots + b_{ic}Row_c(D).$$

Hence each row of A is a linear combination of the rows of D, and so the dimension of the row space of A does not exceed the number of rows of D; that is, $r \leq c$. The same argument may be applied to A^T, so we also have $c =$ row rank of $A^T \leq$ column rank of $A^T = r$. Thus $r = c$.

Dave Stanford
College of William and Mary
Contributed

A Proof of the Equality of
Column and Row Rank of a Matrix

Let A be an $m \times n$ complex matrix, A^* its conjugate transpose, and let x and y denote $n \times 1$ matrices (column vectors).

Lemma 1. $y^*y = 0$ *if and only if* $y = 0$.

Lemma 2. $Ax = 0$ *if and only if* $A^*Ax = 0$.

Proof. If $A^*Ax = 0$, then $y^*y = 0$, where $y = Ax$. Hence, by Lemma 1, $y = 0$. The converse is obvious.

Now let $R(A)$ denote the range of A, i.e., the vector space $\{Ax; \text{ all } x\}$. Note that column rank (c.r.) $A = \dim R(A)$. We write r.r. A for the row rank of A.

Lemma 3. $\dim R(A) = \dim R(A^*A)$.

Proof. By Lemma 2, $Ax_1, \ldots, Ax_k$ are linearly independent if and only if $A^*Ax_1, \ldots,$ A^*Ax_k are linearly independent.

Theorem. c.r. $(A) = $ c.r. $(A)^* = $ r.r. (A).

Proof. c.r. $(A) = $ c.r. $(A^*A) = \dim\{A^*(Ax); \text{ all } x\} \leq \dim\{A^*y; \text{ all } y\} = $ c.r. (A^*). Thus also c.r. $(A^*) \leq $ c.r. $(A^{**}) = $ c.r. (A), and so c.r. $(A) = $ c.r. $(A^*) = $ r.r. $(\overline{A}) = $ r.r. (A). Of course, $\overline{A}$ is the entrywise conjugate of A.

Hans Liebeck
University of Keele
American Mathematical Monthly **73** (1966), 1114

The Frobenius Rank Inequality

Given the product ABC of three rectangular matrices, the Frobenius inequality on rank states that

$$\text{rank } AB + \text{rank } BC \leq \text{rank } B + \text{rank } ABC.$$

Special cases of this inequality provide several familiar facts. For example, $A = 0$ implies rank $BC \leq$ rank B; $C = 0$ implies rank $AB \leq$ rank B; and if B is the $n \times n$ identity matrix, then rank $A +$ rank $C \leq n +$ rank AC. Thus, in particular, Sylvester's inequality is a consequence of the Frobenius inequality. See, for example, [1] or [2].

The proof given below of this inequality is based upon three observations. First, if A is an $m \times n$ matrix over a field F with range $\mathcal{R}(A) = \{AX | X \in F^n\}$, then rank $A = \dim \mathcal{R}(A)$, the dimension of the range as a subspace of F^m. Second, if a linear transformation $T : U \to V$ of dimensional vector space V over F with quotient space V/W of cosets $\nu + W$, then $\dim(V/W) = \dim V - \dim W$.

Proof of the Frobenius inequality. (Compare [3].) Let the product ABC of three matrices over a field be defined. Since $\mathcal{R}(BC) \subseteq \mathcal{R}(B), \mathcal{R}(ABC) \subseteq \mathcal{R}(AB), A(\mathcal{R}(B)) \subseteq \mathcal{R}(AB)$, and $A(\mathcal{R}(BC)) \subseteq \mathcal{R}(ABC)$, then

$$T : \frac{\mathcal{R}(B)}{\mathcal{R}(BC)} \to \frac{\mathcal{R}(AB)}{\mathcal{R}(ABC)}; \quad x + \mathcal{R}(BC) \mapsto Ax + \mathcal{R}(ABC)$$

is a linear transformation. Moreover, since $\mathcal{R}(AB) \subseteq A(\mathcal{R}(B))$, T is onto. Therefore,

$$\begin{aligned}
\text{rank } B - \text{rank } (BC) &= \dim\left(\mathcal{R}(B)/\mathcal{R}(BC)\right) \\
&\geq \dim(\mathcal{R}(AB)/\mathcal{R}(ABC)) \\
&= \text{rank } (AB) - \text{rank } (ABC).
\end{aligned}$$

References

1. R.A. Horn and C.R. Johnson, *Matrix Analysis*, Cambridge University Press, New York, 1985, p. 13.

2. M. Marcus and H. Minc, *A Survey of Matrix Theory and Matrix Inequalities*, Allyn and Bacon, Boston, 1964, pp. 27–28.

3. Donald W. Robinson, A Proof of the Frobenius Rank Equality, *American Mathematical Monthly* **87** (1980), pp. 481–482.

Donald Robinson
Brigham Young University
Contributed

A New Algorithm for Computing the Rank of a Matrix

To determine the rank of a matrix A and a basis for its row space, students are usually instructed to reduce A to echelon form by using elementary row operations. Then the number of nonzero rows of the echelon matrix is the rank of A, and the rows of the echelon matrix constitute a basis for the row space of A.

While this is quite correct, in practice the computations can become unwieldy. For example, if A is an integer matrix there are two standard ways to proceed. In the first approach we start by using elementary row operations (based on the division algorithm) to eliminate all but one entry in the first column of A. In the second approach we first divide the top row by the left-hand entry a_{11} (assuming $a_{11} \neq 0$), then subtract suitable multiples of the new top row from all the others; in either case the computation reduces matters to considering an $(m-1) \times (n-1)$ matrix, whereupon the process is repeated.

The first approach has the advantage that all computations are performed with integers, but a great many row operations are likely to be needed, and there may be some uncertainty as to exactly which row operations should be used in order to minimize computation. In the second approach denominators will immediately appear on the scene (unless A has been specially designed to make computations easy, say by having lots of rows with leading entry 1 involved in the calculations), and the potential for error then increases dramatically. Indeed, many texts make the appearance of nontrivial denominators especially likely by including the requirement that each nonzero row of an "echelon" matrix have 1 as its first nonzero entry. That requirement is irrelevant for rank and row space determinations. (Of course, a leading 1 *is* important if the reduction is being done in order to solve a system of linear equations.)

Our purpose here is to present a rank algorithm in which division is never needed. Like the usual procedures, this one yields an echelon matrix (not necessarily with a leading 1 in its nonzero rows) that is row-equivalent to the original, and whose nonzero rows constitute a basis for the row space of A. But here there will be no uncertainty about how to proceed, and denominators will not arise in the computations.

First we fix some notation. Let F be a field, and let $M_{m \times n}(F)$ denote the set of $m \times n$ matrices over F. For any matrix $A = (a_{ij}) \in M_{m \times n}(F)$ and any pair of indices i, j satisfying $2 \leq i \leq m$ and $2 \leq j \leq n$, we define the 2×2 subdeterminant

$$d_{ij} = \begin{vmatrix} a_{11} & a_{1j} \\ a_{i1} & a_{ij} \end{vmatrix} = a_{11}a_{ij} - a_{i1}a_{1j}.$$

The following result provides the key to our procedure.

175

THEOREM. *Let* $A = (a_{ij}) \in M_{m \times n}(F)$, *and suppose* $a_{11} \neq 0$. *Then*

$$\text{rank} \quad A = 1 + \text{rank} \begin{pmatrix} d_{22} & \cdots & d_{2n} \\ \vdots & & \\ d_{m2} & \cdots & d_{mn} \end{pmatrix}.$$

Proof. Write $\sim$ for row equivalence of matrices. We first multiply rows 2 through m by a_{11}, obtaining

$$A \sim \begin{pmatrix} a_{11} & a_{12} & \cdots & a_{1n} \\ a_{11}a_{21} & a_{11}a_{22} & \cdots & a_{11}a_{2n} \\ \vdots & & & \\ a_{11}a_{m1} & a_{11}a_{m2} & \cdots & a_{11}a_{mn} \end{pmatrix}.$$

Now we take this new matrix and, for each row index i satisfying $2 \leq i \leq m$, we subtract a_{i1} times row 1 from row i. This yields the row equivalence

$$A \sim \begin{pmatrix} a_{11} & a_{12} & \cdots & a_{1n} \\ 0 & d_{22} & \cdots & d_{2n} \\ \vdots & & & \\ 0 & d_{m2} & \cdots & d_{mn} \end{pmatrix}.$$

Because $a_{11} \neq 0$, the first row of this last matrix is not a linear combination of the other rows. The conclusion follows. $\square$

An algorithm for rank computation. First note that if a given nonzero matrix $A = (a_{ij})$ has a leading column of zeros, then the rank is unchanged when that column is deleted; thus we may assume the first column to be nonzero. Moreover, by using an elementary row operation (if necessary) we can assume without loss of generality that $a_{11} \neq 0$. Making use of these observations as needed, we now apply the theorem recursively until we obtain a matrix with only one row or column, and then the solution is evident. (In practice, the rank of a matrix with only two rows or two columns is usually evident.)

Examples.

$$(i) \quad \text{rank} \begin{pmatrix} 3 & -8 & 7 \\ 5 & -4 & 9 \\ 2 & 3 & 6 \end{pmatrix} = 1 + \text{rank} \begin{pmatrix} 28 & -8 \\ 25 & 4 \end{pmatrix} = 2 + \text{rank}(312) = 3.$$

$$(ii) \quad \text{rank} \begin{pmatrix} 4 & 3 & -5 & 6 \\ 6 & 2 & 0 & 2 \\ 3 & 5 & -12 & 5 \\ 2 & 2 & -4 & 2 \end{pmatrix} = 1 + \text{rank} \begin{pmatrix} -10 & 30 & -28 \\ 11 & -33 & 2 \\ 2 & -6 & -4 \end{pmatrix}$$

$$= 2 + \text{rank} \begin{pmatrix} 0 & 288 \\ 0 & 96 \end{pmatrix} = 3.$$

Remarks. (a) The given matrix A is row-equivalent to the echelon matrix constructed by stacking up the top rows of the matrices appearing in the rank computations, with the understanding that leading zeros need to be prefixed to the rows that have fewer than n entries. For example, the matrix A in Example (i) is row-equivalent to the echelon matrix

$$\begin{pmatrix} 3 & -8 & 7 \\ 0 & 28 & -8 \\ 0 & 0 & 312 \end{pmatrix},$$

and the rows of this matrix constitute a basis of the row space of A. (Of course in this example A is nonsingular, so its *own* rows are also a basis for the row space.) Similarly, in Example (ii) the nonzero rows of the echelon matrix

$$\begin{pmatrix} 4 & 3 & -5 & 6 \\ 0 & -10 & 30 & -28 \\ 0 & 0 & 0 & 288 \\ 0 & 0 & 0 & 0 \end{pmatrix}$$

constitute a basis for the row space of A. In building up the echelon matrices in this way, one must be careful to reinsert any columns of zeros that were deleted during the rank computations.

(b) Let A be an $n \times n$ matrix. Each reduction from a $k \times k$ matrix to a $(k-1) \times (k-1)$ matrix requires $2(k-1)^2$ multiplications. Adding these numbers for $k = 2, 3, \ldots, n$, gives a total of $(2n^3 - 3n^2 + n)/3 \approx 2n^3/3$ multiplications, which has the same order of magnitude as the number required to compute the determinant of A (see [1, pp. 479–480]).

Reference

1. Donald E. Knuth, *The Art of Computer Programming*, Volume 2 (second edition), Addison Wesley, Reading, Mass., 1981.

Larry J. Gerstein
University of California, Santa Barbara
American Mathematical Monthly **95** (1988), 950–952

Elementary Row Operations and *LU* Decomposition

The purpose of this capsule is to show how to use the computer as a "matrix calculator," thereby enabling a greater emphasis on concepts rather than arithmetic. Our software is *MATRIXPAD, ver. 2.0*, Morris Orzech, D.C. Heath, Lexington, MA, 1986. We could also implement the lesson with any of several other matrix calculators now available (see [D.P. Kraines and V.Y. Kraines, Linear algebra software for the IBM PC, *College Mathematics Journal* 21 (1990), 56–64]).

At the start of a standard sophomore-level linear algebra course, much class and homework time is spent solving systems of linear equations by Gauss or Gauss-Jordan elimination, using elementary row operations (ERO's). The conceptual issues are often lost in the arithmetic tedium. Most computer matrix packages have an option for finding the row echelon or reduced row echelon form of a matrix. Overreliance on this single-key operation, however, leads to the danger of the "black box" phenomenon; students press keys, record the outputs, and learn nothing. A happy compromise is to have students instruct the computer to do individual ERO's; for example, add -5 times row 2 to row 4 or interchange row 1 and row 3.

MATRIXPAD has an individual ERO option. Its screen layout looks and works like a four-register calculator, with two of the four entries (matrices of sizes up to 8×8) displayed at any time. Each operation, other than entry of a matrix, is a single keystroke; matrices can be stored and recalled from a disk to save class time. *MATRIXPAD* allows both rational (exact) and decimal (approximate) arithmetic. The default mode is rational, which avoids the need to interpret answers such as 1.234 E-12.

Lesson. Discuss the matrix representation $AX = B$ of the system of linear equations

$$3x + 6y - 5z = 0$$
$$4x + 7y - 3z = 9$$
$$3x + 5y - z = 10.$$

Guide the students through the steps for solving such a system: Enter the augmented matrix $H = [A|B]$ (or save time by storing it on the disk before class and recalling it). Activate the ERO mode, and use row operations to reduce H to row echelon form. The resultant system can then be solved by backwards substitution.

Students quickly learn that the program will produce the row echelon form with one keystroke; indeed, you will want to use that mode in class after the first few examples. ERO's will continue to play a significant role, however, when you take up the *LU* decomposition, an important operation in numerical linear algebra that is often neglected in standard courses.

The problem of LU decomposition is to find a lower triangular matrix L with 1's along its diagonal and an upper triangular matrix U (a row echelon form of A) so that $A = LU$. Try to construct U by working from top to bottom, left to right, using only the add operation. If this can be done, i.e., if no row exchanges are necessary, you can construct L in the following way: On the blackboard, make a 3×3 array with 1's along the diagonal and 0's above. Then record in the remaining positions the negative of each rational number multiplier. To check the answer, move U up to the second register of the calculator and enter L from the blackboard. Now exchange the positions of L and U (to make L the left factor), and multiply. Check that the answer is A.

To test students' understanding of ERO's, give them a take-home problem to find the LU decomposition of the 3 by 3 matrix whose entries are the digits of their Social Security numbers.

David P. Kraines
Duke University

Vivian Kraines
Meredith College

David A. Smith
Duke University
College Mathematics Journal **21** (1990), 418–419

A Succinct Notation for Linear Combinations of Abstract Vectors[*]

A succinct, uncluttered notation for expressing linear combinations of abstract vectors is presented. The notation is not to be confused with matrices.

1. Matrix background

Let $\mathbf{y}$ be an m-tuple, i.e. an $m \times 1$ matrix. Let $\mathbf{A}$ be an $m \times n$ matrix. Let $\mathbf{x}$ be an n-tuple. Let the columns of $\mathbf{A}$ be denoted by $\mathbf{a}_1, \mathbf{a}_2, \ldots, \mathbf{a}_n$. Using matrix partitioning,

$$\mathbf{A} = (\mathbf{a}_1, \mathbf{a}_2, \ldots, \mathbf{a}_n) \tag{1}$$

The equation

$$\mathbf{y} = \mathbf{A}\mathbf{x} \tag{2}$$

has several interpretations. The interpretation which will concern us in this paper is: $\mathbf{y}$ is a linear combination of the columns of $\mathbf{A}$; this may also be expressed by

$$\mathbf{y} = \sum_{i=1}^{n} x_i \mathbf{a}_i \tag{3}$$

However, (2) is a more succinct and less cluttered notation than (3).

2. The notation

An abstract vector space satisfies eight well known axioms. (See, for example, [1, pp. 6–7].)

Let

$$(\mathbf{v}_1 \mathbf{v}_2 \cdots \mathbf{v}_n) \tag{4}$$

be an ordered set of vectors in a given abstract vector space. (An n-tuple, of course, is a special case of an abstract vector.) Let $\mathbf{V}$ be a succinct notational equivalent for (4). Thus

$$\mathbf{V} = (\mathbf{v}_1 \mathbf{v}_2 \cdots \mathbf{v}_n) \tag{5}$$

---

[*] "A Succinct Notation for Linear Combinations of Abstract Vectors" by Leon Katz, *International Journal of Mathematical Education in Science and Technology*, vol. 18 (1977) pp. 47–50. Reprinted with permission by Taylor and Francis Ltd. http://www.tandf.co.uk/journals.

I intend (5) to have a formal resemblance to (1). Let

$$\mathbf{c} = \begin{pmatrix} c_1 \\ c_2 \\ \vdots \\ c_n \end{pmatrix}$$

be an n-tuple.

Consider the linear combination of abstract vectors given by

$$\mathbf{w} = \sum_{i=1}^{n} c_i \mathbf{v}_i \tag{6}$$

This brings us to the crux of the matter. Let

$$\mathbf{w} = \mathbf{V}\mathbf{c} \tag{7}$$

be a succinct way of writing the linear combination of the abstract vectors. Thus, by definition (7) and (6) are equivalent. I intend, of course, (7) to have a formal resemblance to (2). I trust that the reader will appreciate the succinctness and freedom from clutter of (7) as compared with (6).

Provided one keeps in mind that $\mathbf{V}$ is *not* a matrix, there is a pedagogic and mnemonic advantage to referring to $\mathbf{v}_1, \mathbf{v}_2, \ldots, \mathbf{v}_n$ as the 'columns' of $\mathbf{V}$.

It is useful to have a name for notation such as $\mathbf{V}$. Let us call such a notation a *matrixlike*.

It is only one step further to write an equation such as

$$\mathbf{W} = \mathbf{V}\mathbf{A} \tag{8}$$

where $\mathbf{W}$ is a matrixlike having n 'columns', $\mathbf{V}$ is a matrixlike having m 'columns', and $\mathbf{A}$ is an $m \times n$ matrix. For $j = 1, 2, \ldots, n$, the jth 'column' of $\mathbf{W}$ is a linear combination of the 'columns' of $\mathbf{V}$; the jth column of $\mathbf{A}$ contains the coefficients used in this linear combination. Thus, (8) is a very compact and uncluttered notation for denoting n linear combinations of abstract vectors. It can be used in lieu of writing several equations which contain '$\sum$' notation.

Now consider an abstract vector space with basis

$$\begin{pmatrix} \mathbf{v}_1 \mathbf{v}_2 \cdots \mathbf{v}_n \end{pmatrix}$$

Let

$$\mathbf{V} = \begin{pmatrix} \mathbf{v}_1 \mathbf{v}_2 \cdots \mathbf{v}_n \end{pmatrix}$$

Then $\mathbf{V}$ is called a *basis matrixlike*. Consider an arbitrary abstract vector in the vector space; let $\mathbf{x}$ be the n-tuple comprised of the coordinates of the arbitrary vector with respect to the 'columns' of $\mathbf{V}$. Then the arbitrary abstract vector, itself, is given simply by $\mathbf{V}\mathbf{x}$.

3. Applications

3.1 *Linear transformations*

Let u-space be an abstract vector space and let the 'columns' of $\mathbf{U}$ be a basis for u-space. Let v-space be another abstract vector space and let the 'columns' of $\mathbf{V}$ be a basis for v-space. Let $\mathcal{T}$ be a linear transformation from u-space into v-space. For $j = 1, 2, \ldots, n$ let the jth column of matrix $\mathbf{A}$ contain the coordinates with respect to the basis in v-space of the image under the transformation of the jth basis vector for u-space. Then the relationships are clearly exhibited by using the matrixlike notation

$$\mathcal{T}\mathbf{u}_j = \mathbf{V}\mathbf{a}_j$$

The n relationships can be collectively expressed by the equation

$$\mathcal{T}\mathbf{U} = \mathbf{V}\mathbf{A} \tag{9}$$

Let $\mathbf{x}$ be the n-tuple comprised of the coordinates of a given arbitrary vector in u-space (with respect to the basis for u-space). Postmultiply (9) by $\mathbf{x}$ to obtain

$$\mathcal{T}\mathbf{U}\mathbf{x} = \mathbf{V}\mathbf{A}\mathbf{x} \tag{10}$$

which says that the coordinates (with respect to the basis for v-space) of the image of the given arbitrary vector are given by $\mathbf{A}\mathbf{x}$. (I have implicitly used the equality $(\mathbf{V}\mathbf{A})\mathbf{x} = \mathbf{V}(\mathbf{A}\mathbf{x})$, which can be proved without difficulty by exploiting the formal resemblance to matrix multiplication.) Equations (9) and (10) give the essence of the role of matrices in linear transformations.

3.2. *Change of basis*

Let the 'columns' of $\mathbf{W}$ be a basis for an abstract vector space and let the 'columns' of $\mathbf{V}$ be another basis for the same space. Let $\mathbf{A}$ be the matrix which gives the coordinates of the vectors in one basis with respect to the other basis; specifically,

$$\mathbf{W} = \mathbf{V}\mathbf{A} \tag{11}$$

Let $\mathbf{x}$ be the n-tuple comprised of the coordinates of a given arbitrary vector with respect to the basis given by the 'columns' of $\mathbf{W}$. Postmultiply (11) by $\mathbf{x}$ to obtain

$$\mathbf{W}\mathbf{x} = \mathbf{V}\mathbf{A}\mathbf{x} \tag{12}$$

which says that the n-tuple $\mathbf{A}\mathbf{x}$ contains the coordinates of the given arbitrary vector with respect to the basis given by the 'columns' of $\mathbf{V}$. Equations (11) and (12) give the essence of the role of matrices in change of basis.

3.3. Mystery theorem

As a challenge to the reader—and to illustrate the power of the notation—a well known theorem concerning abstract vectors is given below. Only the essence of the hypotheses and only the essence of the proof are given.

THEOREM (condensed)

1. $\mathcal{T}\mathbf{V} = \mathbf{VA}$
2. $\mathcal{T}\mathbf{W} = \mathbf{WB}$
3. $\mathbf{W} = \mathbf{VC}$

$\rightarrow$ $\mathbf{B} = \mathbf{C}^{-1}\mathbf{AC}$

PROOF (condensed)

$\mathcal{T}\mathbf{VC} = \mathbf{VAC}$ (by Hypothesis 1)
$\mathcal{T}\mathbf{VC} = \mathbf{VCB}$ (by Hypotheses 2 and 3)
$0 = \mathbf{V}(\mathbf{AC} - \mathbf{CB})$ (by subtraction)
$0 = \mathbf{AC} - \mathbf{CB}$ (by linear independence)
$\mathbf{CB} = \mathbf{AC}$
$\mathbf{B} = \mathbf{C}^{-1}\mathbf{AC}$ Q.E.D.

The challenge for the reader is to interpret the hypotheses.

4. Motivation

The matrixlike notation enables us to write and manipulate very simple equations which express the non-simple relationships of matrices to abstract vector spaces.

I confess to having experienced extreme difficulty in understanding the relationships of matrices to abstract vector spaces. All explanations seem to contain verbose English sentences and numerous, elusive equations containing cluttered '$\sum$' summations. I always get lost. Expressing linear combinations of abstract vectors in a manner which parallels that of n-tuples seemed natural and led me to formulate the matrixlike notation. It was only through use of the notation that I was able to comprehend the relationships of matrices to abstract vector spaces. I trust that the matrixlike notation will prove beneficial to others.

5. Printing/handwriting suggestion

It may sometimes be helpful (although I have not shown it here) to use a special underlining to distinguish the appearance of abstract vectors from n-tuples and to distinguish the appearance of matrixlikes from matrices. Ordinary underlining is not recommended because it is often used to indicate n-tuples.

Reference

1. Kreider, D.L., Kuller, R.G., Ostberg, D.R., and Perkins, F.W., 1966, *An Introduction to Linear Analysis* (Reading, Massachusetts: Addison-Wesley).

Leon Katz
Arlington, Virginia
Int. J. Math. Educ. Sci. Tech. **18** (1987), 47–50

Why Should We Pivot in Gaussian Elimination?

The Gaussian elimination procedure for solving a system of n linear equations in n unknowns is familiar to most precalculus students: (1) Write the system as an augmented matrix, (2) reduce the system to upper triangular form by the elementary row operations, and (3) solve for the variables by back substitution. The method is simple and terminates in a finite number of steps, with either the exact answer or the information that there is no unique solution. This straightforward procedure seems ideal for computer implementation. As long as we pivot (that is, interchange rows) to avoid division by zero, what can possibly go wrong? Well, almost everything can go wrong, as illustrated by the following examples.

Example 1. Consider the matrix equation

$$\begin{bmatrix} \epsilon & 1 & 1 \\ 1 & -1 & 1 \\ .5 & 1 & 1 \end{bmatrix} \begin{bmatrix} x_1 \\ x_2 \\ x_3 \end{bmatrix} = \begin{bmatrix} 2 \\ 1 \\ 2.5 \end{bmatrix},$$

where ϵ is a constant. When *epsilon* $= 0$, a standard "naive" Gaussian elimination algorithm, which pivots only to avoid division by zero, will yield the exact solution $(x_1, x_2, x_3) = (1, 1, 1)$. When ϵ is positive but much smaller than 0.5, the answer should remain near to this solution: this system is "well-conditioned" for such ϵ in the sense that small changes in the coefficients give rise to small changes in the answer. Indeed, we see this from the exact solution

$$\left(x_1, x_2, x_3 \right) = \left(1 + \frac{4\epsilon}{2 - 4\epsilon}, 1 + \frac{\epsilon}{2 - 4\epsilon}, 1 - \frac{3\epsilon}{2 - 4\epsilon} \right). \tag{20}$$

However, an implementation of the algorithm on an Apple Macintosh in the binary version of Microsoft Basic (which stores real numbers with 24 bit mantissas—about seven significant decimal digits) produced the following results for small values of ϵ:

$\epsilon = 2^{-23}$ $(x_1, x_2, x_3) = (1.000\,000, 0.999\,9999, 1.000\,000)$
$\epsilon = 2^{-23.5}$ $(x_1, x_2, x_3) = (2.828\,427, 0.999\,9998, 1.000\,000)$
$\epsilon = 2^{-24}$ $(x_1, x_2, x_3) = (0.000\,000, 2.000\,000, 0.000\,000)$
$\epsilon = 2^{-25}$ $(x_1, x_2, x_3) = $ doesn't exist, as the coefficient matrix is computed
                                            to be singular!

Notice that for $\epsilon \approx 0$, the coefficient matrix is not even close to being singular, since the system is well-conditioned. In fact, for $\epsilon = 0$, the coefficient matrix and its inverse are given by

$$A = \begin{bmatrix} 0 & 1 & 1 \\ 1 & -1 & 1 \\ .5 & 1 & 1 \end{bmatrix} \quad \text{and} \quad A^{-1} = \begin{bmatrix} -1 & 0 & 2 \\ -0.5 & -0.5 & 1 \\ 1.5 & 0.5 & -1 \end{bmatrix}.$$

Very different and even more startling things happen to the *exact* solution when ϵ is close to 0.5, for then the coefficient matrix is nearly singular. (If $\epsilon = 0.5$, then the first and third rows of our original matrix are identical.) In order to analyze this case, let $\epsilon = 0.5 + \delta$, where δ is a small nonzero constant. Then the exact solution of our matrix equation is

$$(x_1, x_2, x_3) = \left(-\frac{1}{2\delta}, \ -\frac{1}{8\delta} + \frac{3}{4}, \ \frac{3}{8\delta} + \frac{7}{4} \right). \tag{21}$$

Observe, for example, the *exact* solution corresponding to each of the values δ:

$$\delta = 2^{-24} \quad (x_1, x_2, x_3) = (-8\,388\,608, -2\,097\,151.25, 6\,291\,457.75)$$

$$\delta = -2^{-24} \quad (x_1, x_2, x_3) = (8\,388\,608, 2\,097\,152.75, -6\,291\,454.25).$$

These spectacular changes in the exact solution are typical of ill-conditioned (that is, non well-conditioned) systems. Surprisingly, the same program that performed so erratically for $\epsilon \approx 0$ performed beautifully for $\epsilon \approx .5$, as we will see in Table 2.

The tremendous changes in the exact solution can be explained geometrically in terms of the "tipping" of one of the two nearly parallel planes given by the first and third equations: Think of holding a long board on the palms of your hands just above a desktop. If you drop your right hand down slightly, the plane of the board will intersect the plane of the desktop to the right of the desktop. Then if you drop your other hand down, the line of the intersection will move to the other side of the desktop. A small change in the balancing of the board results in a large change in the line of the intersection between the planes of the board and the desktop. In the well-conditioned case considered earlier, these two planes intersect at an angle of about 20°. The figures illustrate the situation in two dimensions.

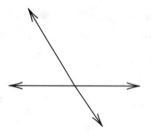

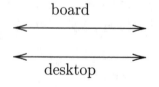

A well-conditioned system: a small change in the coefficients produces a small change in the point of intersection.

An ill-conditioned system: a small change in the coefficients (tipping the board down on the left or right slightly) produces a large change in the point of intersection.

The well-conditioned case, $\epsilon \approx 0$. What can account for the wide diversity of incorrect answers we see when $\epsilon \approx 0$? An obvious answer is, "round-off errors caused by computer arithmetic." Although there are such errors, this does not adequately explain our solution results nor does it help us to alleviate the difficulties. A less obvious, but correct, explanation is that when we used the first equation to eliminate x_1 from the second and third equations, we divided by a small, nearly zero "pivot" (namely, ϵ). Unfortunately, many numerical analysis textbooks go no further than making this observation. There are at least three compelling reasons to pursue this observation further. Not doing so (i) leaves students with the impression that dividing by a small number, in and of itself, causes roundoff errors, (ii) it does not adequately account for the results above, and (iii) stopping at this point misses a valuable opportunity to underline the most common causes of roundoff errors and the interplay between them—namely, the subtraction of almost equal quantities (the most common source of devastating roundoff errors) and the addition of a small, important number to a large, relatively unimportant number.

Our intent here is to pursue this observation and demonstrate how small pivots may result in:

1. loss of significant figures due to subtraction of almost equal quantities during the back substitution process,

2. loss of significant figures due to the addition of large and small quantities and subsequent subtraction of almost equal quantities during the forward elimination procedure.

The first of these problems can be easily illustrated by the following.

Example 2. Solve

$$\begin{bmatrix} \epsilon & 1 \\ 1 & -1 \end{bmatrix} \begin{bmatrix} x_1 \\ x_2 \end{bmatrix} = \begin{bmatrix} 1 \\ 0 \end{bmatrix}.$$

Naive Gaussian elimination tells us to replace row 2 by row (row 2)$-1/\epsilon$ (row 1), yielding

$$\left[\begin{array}{cc|c} \epsilon & 1 & 1 \\ 1 & -1 & 0 \end{array} \right] \rightarrow \left[\begin{array}{cc|c} \epsilon & 1 & 1 \\ 0 & -1-\dfrac{1}{\epsilon} & -\dfrac{1}{\epsilon} \end{array} \right].$$

If $1/\epsilon$ is so large that the computer replaces $-1 - (1/\epsilon)$ by $-1/\epsilon$, then x_2 will be assigned the value 1. This is quite accurate (taking $\epsilon = 10^{-8}$, for example, we see that $x_2 = -10^8/(-1 - 10^8) = .999\,999\,990\ldots$.) Note, however, that the assignment $x_2 = 1$ results in the computation $x_1 = 0$. This is very bad since, in fact, $x_1 = x_2$; there are no significant figures left in the computed solution. The loss of significance occurred during the back substitution process when almost equal quantities (namely, 1 and x_2) were subtracted.

Let's consider the general solution of

$$\begin{bmatrix} a_{11} & a_{12} & \cdots & a_{1n} \\ & a_{22} & \cdots & a_{2n} \\ & & \mathbf{0} \quad \ddots & \vdots \\ & & & a_{nn} \end{bmatrix} \begin{bmatrix} x_1 \\ x_2 \\ \vdots \\ x_n \end{bmatrix} = \begin{bmatrix} b_1 \\ b_2 \\ \vdots \\ b_n \end{bmatrix},$$

where the reduced upper triangular matrix has been obtained by elimination. Then the back substitution formula for x_i is

$$x_i = \frac{b_i - \sum\limits_{j=i+1}^{n} a_{ij}x_j}{a_{ii}},$$

where the $x_j \, (i+1 \le j \le n)$ have already been computed by back substitution. If a_{ii} is small, then its reciprocal is large. If x_i is *not* large, then the numerator, $b_i - \sum_{j=i+1}^{n} a_{ij}x_j$, must be small. Finally, if b_i is not small, then it must follow that $\sum_{j=i+1}^{n} a_{ij}x_j$ and b_i must be nearly equal. Therefore, a small pivot a_{ii} may result in the loss of significant digits due to subtraction of almost equal quantities during the back substitution process. This is what occurred in Example 2, for $i = 1$.

Remark. There is more than one way to compute x_i. We may first compute $\sum_{j=1+i}^{n} a_{ij}x_j$ and then subtract it from b_i, as suggested above. We may also start with b_i and subtract each of the terms $a_{ij}x_j$, one at a time. The author obtained the same results by both methods, but this may vary somewhat on different computers.

Example 2 continued. If we interchange the first two rows of the original augmented matrix, the pivot will no longer be small, and subtraction of almost equal quantities will no longer occur:

$$\left[\begin{array}{cc|c} 1 & -1 & 0 \\ \epsilon & 1 & 1 \end{array} \right] \rightarrow \left[\begin{array}{cc|c} 1 & -1 & 0 \\ 0 & 1+\epsilon & 1 \end{array} \right].$$

Then back substitution yields

$$x_2 = \frac{1}{1+\epsilon} \approx 1 \text{ if } \epsilon \text{ is very small,}$$

and

$$x_1 = x_2 \approx 1.$$

To see what may go wrong during the elimination phase of the algorithm, we need an example with three equations and three unknowns.

Example 1 revisited. If we apply naive Gaussian elimination to the augmented matrix in Example 1, we get

$$\left[\begin{array}{ccc|c} \epsilon & 1 & 1 & 2 \\ 0 & -1-\dfrac{1}{\epsilon} & 1-\dfrac{1}{\epsilon} & 1-\dfrac{2}{\epsilon} \\ 0 & 1-\dfrac{.5}{\epsilon} & 1-\dfrac{.5}{\epsilon} & 2.5-\dfrac{1}{\epsilon} \end{array} \right]. \tag{22}$$

If $1/\epsilon$ is large, then all the entries with a division by ϵ involve the subtraction of a large number from a (relatively) small number. If the computer replaces all these entries by the larger number, the resulting matrix is

$$\begin{bmatrix} \epsilon & 1 & 1 & \Big| & 2 \\ 0 & -\dfrac{1}{\epsilon} & -\dfrac{1}{\epsilon} & \Big| & -\dfrac{2}{\epsilon} \\ 0 & -\dfrac{.5}{\epsilon} & -\dfrac{.5}{\epsilon} & \Big| & -\dfrac{1}{\epsilon} \end{bmatrix}.$$

Applying elimination one more time yields the matrix:

$$\begin{bmatrix} \epsilon & 1 & 1 & \Big| & 2 \\ 0 & -\dfrac{1}{\epsilon} & -\dfrac{1}{\epsilon} & \Big| & -\dfrac{2}{\epsilon} \\ 0 & 0 & 0 & \Big| & 0 \end{bmatrix},$$

so the computer reports that there is no unique solution. This is evidently what was observed earlier with $\epsilon = 2^{-25}$.

If the computer does not replace entries such as $-1 - (1/\epsilon)$ by $-1/\epsilon$, then continuing the elimination from the augmented matrix labeled (3) would yield:

$$\begin{bmatrix} \epsilon & 1 & 1 & \Big| & 2 \\ 0 & -1-\dfrac{1}{\epsilon} & 1-\dfrac{1}{\epsilon} & \Big| & 1-\dfrac{2}{\epsilon} \\ 0 & 0 & \left(1-\dfrac{.5}{\epsilon}\right) - \left[\dfrac{1-\dfrac{.5}{\epsilon}}{-1-\dfrac{1}{\epsilon}}\right]\left(1-\dfrac{1}{\epsilon}\right) & \Big| & \left(2.5-\dfrac{1}{\epsilon}\right)-\left[\dfrac{1-\dfrac{.5}{\epsilon}}{-1-\dfrac{1}{\epsilon}}\right]\left(1-\dfrac{2}{\epsilon}\right) \end{bmatrix}.$$

The computations in the third row may result in a loss of significant digits due to the subtraction of almost equal quantities. There are at least two possibilities:

(i) x_3 is computed with fewer significant digits than desired, and consequently the computations of x_2 and x_1 during the back substitution process also have fewer digits than desired. When we let $\epsilon = 2^{-24}$, the last entry in column four was computed to be zero, and the resulting computations were

$$x_3 = 0, \quad x_2 = \frac{-2/\epsilon}{-1/\epsilon} = 2, \quad \text{and} \quad x_1 = 0.$$

(ii) x_3 is computed as nearly 1, and x_2 is also computed as nearly 1. However, x_1 is computed by the back substitution formula, $x_1 = (2 - x_2 - x_3)/\epsilon$, to be far from 1 because of the subtraction of almost equal quantities. When we let $\epsilon = 2^{-23.5}$, this is exactly what happened. (Curiously, x_1 was computed to be $2\sqrt{2}$; perhaps $2 - x_2 - x_3$ was computed to be 2^{-22} rather than $2^{-23.5}$.) Here we are seeing the same back substitution phenomena as illustrated in Example 2.

The strategy that should be employed is now fairly evident:

Before eliminating below a diagonal element, first interchange (if necessary) the row containing the diagonal element with a row below it that will make the absolute value of the diagonal element as large as possible.

This will greatly reduce the number of times nearly equal quantities are subtracted. This commonly used strategy is called "partial (or column) pivoting." ("Complete pivoting" involves both row and column interchanges to secure the pivot with maximal absolute value among *all* remaining candidates. This is much less frequently used.)

Let's illustrate partial pivoting for Example 1. Since 1 is the largest entry in the first column, we interchange rows and 1 and 2 in the original augmented matrix to obtain

$$\left[\begin{array}{ccc|c} 1 & -1 & 1 & 1 \\ \epsilon & 1 & 1 & 2 \\ .5 & 1 & 1 & 2.5 \end{array}\right].$$

Applying elimination yields

$$\left[\begin{array}{ccc|c} 1 & -1 & 1 & 1 \\ 0 & 1+\epsilon & 1-\epsilon & 2-\epsilon \\ 0 & 1.5 & .5 & 2 \end{array}\right].$$

Since ϵ is small, continuing the elimination process will not result in subtractions of almost equal quantities. In particular, if ϵ is so small that the computer drops it, the augmented matrix becomes

$$\left[\begin{array}{ccc|c} 1 & -1 & 1 & 1 \\ 0 & 1 & 1 & 2 \\ 0 & 1.5 & .5 & 2 \end{array}\right]$$

Since 1.5 is the largest entry in column 2 on or below the diagonal, partial pivoting requires us to interchange rows 2 and 3. This gives

$$\left[\begin{array}{ccc|c} 1 & -1 & 1 & 1 \\ 0 & 1.5 & .5 & 2 \\ 0 & 1 & 1 & 2 \end{array}\right].$$

Assume, for simplicity, that the computer uses seven decimal digits for the mantissas. Then the elimination will yield

$$\left[\begin{array}{ccc|c} 1 & -1 & 1 & 1 \\ 0 & 1.5 & .5 & 2 \\ 0 & 0 & .6666667 & .6666667 \end{array}\right].$$

Therefore, back substitution now gives $x_3 = 1, x_2 = 1$, and $x_1 = 1$, all of which are close to true answers given in equation (1).

Computer Experiments

The author coded a fairly standard algorithm for Gaussian elimination on an Apple Macintosh in Microsoft's binary version of Basic. (The algorithm was adapted from the Fortran in [1,pp. 220–223].) The results are summarized in the following tables.

Example 1 one more time. All of the different types of behavior described above were observed for different values of ϵ. In Table 1, the values obtained with pivoting are correct to the seven digits shown (in each case); the error reported is just the difference between the results obtained with pivoting and those obtained without pivoting. The table is in order of decreasing ϵ, so the worst case is last.

Table 1. The well-conditioned case: $\epsilon \approx 0$

		Without pivoting	With pivoting	Error w/o pivoting
$\epsilon = 2^{-5}$	x_1	1.066 666	1.066 667	.000 001
	x_2	1.016 666	1.016 667	.000 001
	x_3	0.950 0005	0.950 0000	−.000 0005
$\epsilon = 2^{-13}$	x_1	1.000 000	1.000 244	.000 244
	x_2	0.999 8779	1.000 061	.000 1831
	x_3	1.000 000	0.999 8168	−.000 1832
$\epsilon = 2^{-23.5}$	x_1	2.828 427	1.000 000	−1.828 427
	x_2	0.999 9998	1.000 000	.000 0002
	x_3	1.000 000	0.999 9997	−0.000 0003
$\epsilon = 2^{-24}$	x_1	0.000 000	1.000 000	1.000 000
	x_2	2.000 000	1.000 000	−1.000 000
	x_3	0.000 000	1.000 000	1.000 000
$\epsilon = 2^{-25}$	x_1	no solution	1.000 000	—
	x_2	no solution	1.000 000	—
	x_3	no solution	1.000 000	—

The ill-conditioned case, $\epsilon \approx .5$. Ill-conditioned systems are a lot tougher to handle than well-conditioned ones. However, since the coefficient matrix of an ill-conditioned system is nearly singular, it makes sense to believe that no matter what pivoting strategy is employed, all the coefficient entries in the last row of the final augmented matrix will be nearly zero. Hence a small pivot will be difficult to avoid, and roundoff errors may easily destroy any confidence we have in the computed solution.

To see what happens in the ill-conditioned case, we ran the program using $\delta = 2^{-24}$ and $\delta = -2^{-24}$ (recall that $\epsilon = .5 + \delta$). These numbers were chosen so that ϵ could be entered without roundoff.

Examination of the numbers in Table 2 shows a remarkable result:

The usual pivoting strategy gave much poorer results!

Table 2. The ill-conditioned case: $\epsilon \approx .5$

		Without pivoting	With pivoting	True solution
$\delta = 2^{-24}$	x_1	$-8,388,608$	$-11,184,810$	$-8,388,608$
	x_2	$-2,097,151$	$-2,796,201$	$-2,097,151.25$
	x_3	$6,291,457$	$8,388,608$	$6,291,457.75$
$\delta = -2^{-24}$	x_1	$8,388,608$	$11,184,810$	$8,388,608$
	x_2	$2,097,153$	$2,796,204$	$2,097,152.75$
	x_3	$-6,291,455$	$- 8,388,608$	$-6,291,454.25$

The reason for this result is that interchanging rows 1 and 2 and doing the elimination yields

$$\left[\begin{array}{ccc|c} 1 & -1 & 1 & 1 \\ 0 & 1.5 + \delta & .5 - \delta & 1.5 - \delta \\ 0 & 1.5 & .5 & 2 \end{array}\right].$$

Now the 2×2 system in the bottom two rows is very nearly singular; continuing the elimination procedure results in the loss of significant digits due to the subtraction of almost equal quantities.

On the other hand, solving the system without pivoting yields the matrix (3) in Example 1, after elimination in the first column:

$$\left[\begin{array}{ccc|c} \epsilon & 1 & 1 & 2 \\ 0 & -1 - \dfrac{1}{\epsilon} & 1 - \dfrac{1}{\epsilon} & 1 - \dfrac{2}{\epsilon} \\ 0 & 1 - \dfrac{.5}{\epsilon} & 1 - \dfrac{.5}{\epsilon} & 2.5 - \dfrac{1}{\epsilon} \end{array}\right] \tag{3}$$

As $\epsilon = .5 + \delta \approx .5$, the second row is computed without loss of significance. Although the third row involves the subtraction of almost equal quantities, hardly any loss of significance occurs since

$$\frac{.5}{\epsilon} = \frac{1}{1 + 2\delta} = 1 - 2\delta + 4\delta^2 - \cdots \approx 1 - 2\delta,$$

yielding

$$1 - \frac{.5}{\epsilon} \approx 2\delta.$$

This is very close to the actual value of $\delta/(.5 + \delta)$. Similarly, the other entries can be written in terms of δ yielding

$$\left[\begin{array}{ccc|c} .5 + \delta & 1 & 1 & 2 \\ 0 & -3 + 4\delta & -1 + 4\delta & -3 + \delta \\ 0 & 2\delta & 2\delta & .5 + 4\delta \end{array}\right].$$

The 2×2 system in the bottom two rows is no longer ill-conditioned since the two lines in the $x_2 x_3$-plane are not nearly parallel. In order to continue the elimination and to model the behavior of a computer, third-order terms will be omitted and, in each computation below only the two lowest-order terms will be retained; that is, $a + b\delta + c\delta^2$ will be replaced by $a + b\delta$ and $(a/\delta) + b + c\delta$ will be replaced by $(a/\delta) + b$. Let's continue the elimination, replacing row 3 by

$$(\text{row 3}) - \frac{2\delta}{-3 + 4\delta}(\text{row 2}) = (\text{row 3}) - \frac{2\delta}{-3}\left(\frac{1}{1 - \dfrac{4\delta}{3}}\right)(\text{row 2})$$

$$\approx (\text{row 3}) + \frac{2\delta}{3}\left(1 + \frac{4\delta}{3}\right)(\text{row 2})$$

$$= (\text{row 3}) + \left(\frac{2\delta}{3} + \frac{8\delta^2}{9}\right)(\text{row 2})$$

to obtain

$$\begin{bmatrix} .5 + \delta & 1 & 1 & 2 \\ 0 & -3 + 4\delta & -1 + 4\delta & -3 + 8\delta \\ 0 & 0 & \dfrac{4\delta}{3} + \dfrac{16\delta^2}{9} & .5 + 2\delta \end{bmatrix}.$$

Back substitution gives

$$x_3 = \frac{1}{\dfrac{4\delta}{3} + \dfrac{16\delta^2}{9}}(.5 + 2\delta) \approx \frac{3}{8\delta} + 1$$

$$x_2 = \frac{-3 + 8\delta + (1 - 4\delta)x_3}{-3 + 4\delta} \approx \frac{1}{8\delta} + 1$$

$$x_1 = \frac{2 - x_2 - x_3}{.5 + \delta} \approx \frac{1}{2\delta} + 1,$$

which are in remarkable agreement with both the exact answer given in equation (2) and the computed answer given in Table 2.

Should we conclude from this example that we shouldn't pivot when solving ill-conditioned problems? No! For if we had *started* with the first two rows interchanged and then solved the system with or without pivoting, we would get exactly the same poor results. We can only conclude that ill-conditioned systems are hard to solve accurately with finite precision and a fixed strategy.

Postscript. Proper use of partial pivoting requires that the coefficients in different rows be somewhat comparable in size. For example, if we multiply the first equation in Example 2 by $1/\epsilon$, then we would not be required to pivot, since each entry in the first column would be one. However, if we were to carry out the elimination and back substitution on the resulting system, we'd observe again that $x_2 = 1$ and $x_1 = 0$. One should start Gaussian elimination with a division of each row by the coefficient in that row which is largest in absolute value. This procedure is called *scaling*, and it makes the absolute value of the largest entry in each row equal to 1. Our examples were chosen so that scaling was unnecessary.

The classic treatment of rounding errors is by J.H. Wilkinson [3]. A thorough treatment of matrix computation can be found in [2].

Acknowledgement. This paper was written while on partial release time provided by the UTC Department of Mathematics and with equipment partially supplied by the UTC Center of Excellence for Computer Applications.

References

1. Ward Cheney and David Kincaid, *Numerical Mathematics and Computing*, Second Edition, Brooks/Cole Publishing Company, Belmont, CA, 1985.
2. Gene H. Golub and Charles F. Van Loan, *Matrix Computations*, The Johns Hopkins University Press, Baltimore, MD, 1983.
3. James H. Wilkinson, *Rounding Errors in Algebraic Processes*, Prentice Hall, Englewood Cliffs, NJ, 1963.

Edward Rozema
University of Tennessee, Chattanooga
College Mathematics Journal **19** (1988), 63–72

The Gaussian Algorithm for Linear Systems with Interval Data

Introduction

The best known method for the solution of linear systems of simultaneous equations is the Gaussian algorithm. Using the fact that the set of real numbers form a field, the given system is first transformed to an equivalent system (that is, to a system that has the same solution) with an upper triangular form. This system can easily be solved for the unknowns by using again the fact that the reals form a field. Subsequently we demonstrate that the Gaussian algorithm can also be used to *enclose* the solution set of a linear system for which the elements of the coefficient matrix and of the right-hand side are allowed to vary independently of each other in given real compact intervals. This holds despite the fact that the set of real compact intervals endowed with the so-called interval operations does not form a field.

All results of this presentation are well known and can be found in the books [2],[6], and [7], for example. The discussion is on an elementary level. Therefore the results can be taught in a course on elementary linear algebra nearly simultaneously with the "normal" Gaussian algorithm.

Interval Operations

We denote the set of real numbers as usual by $\mathbf{R}$ and its elements by $a, b, \ldots, x, y, z$. If $x_1, x_2 \in \mathbf{R}, x_1 \leq x_2$, then the set $[x_1, x_2] = \{x | x_1 \leq x \leq x_2\}$ is called a real compact interval. For ease of notation or if the explicit notation of the bounds is not essential, we write $[x]$ instead of $[x_1, x_2]$.

The set of real compact intervals is denoted by $\mathbf{IR}$, its elements by $[a], [b], \ldots, [x], \ldots$. In $\mathbf{IR}$ we define four basic operations (the so-called interval operations), namely addition, subtraction, multiplication and division, by

$$[x] * [y] = \{x * y | x \in [x], y \in [y]\}, [x], [y] \in \mathbf{IR}, * \in \{+, -, \cdot, /\}. \tag{1}$$

We assume that $0 \notin [y]$ in the case of division. When applying these operations it is important that the result $[x] * [y]$ is again in $\mathbf{IR}$ and that its bounds can be expressed by the bounds of $[x]$ and $[y]$. Let $[x] = [x_1, x_2]$ and $[y] = [y_1, y_2]$. Then we have the rules:

$$[x] + [y] = [x_1 + y_1, x_2 + y_2]$$
$$[x] - [y] = [x_1 - y_2, x_2 - y_1]$$
$$[x] \cdot [y] = [\min\{x_1 y_1, x_1 y_2, x_2 y_1, x_2 y_2\}, \max\{\text{same products}\}]$$
$$[x]/[y] = \left[\min\left\{\frac{x_1}{y_1}, \frac{x_1}{y_2}, \frac{x_2}{y_1}, \frac{x_2}{y_2}\right\}, \max\{\text{same quotients}\}\right], \quad 0 \notin [y].$$

$\mathbf{R}$ is considered as the subset of elements of $\mathbf{IR}$ for which the lower and the upper bounds coincide. The four operations for reals coincide with the interval operations if $\mathbf{R}$ is considered as a subset of $\mathbf{IR}$ endowed with the interval operations. It is a basic fact that the reals $\mathbf{R}$ endowed with the operations of addition and multiplication form a field; That is, with respect to addition, $\mathbf{R}$ forms an abelian group; with respect to multiplication, $\mathbf{R}\backslash 0$ forms an abelian group, and the distributive law holds. Subtraction and division are then introduced via the group properties. The unique solution of the equation $b + x = 0$ is denoted by $x = -b$ and the subtraction $a - b$ is defined to be $a + (-b)$. Similarly if $b \neq 0$ then $1/b$ is the unique solution of the equation $bx = 1$ and the division a/b is defined to be $a \cdot (1/b)$.

In the case of $\mathbf{IR}$ we cannot proceed in this manner. Consider first the equation $[b] + [x] = 0$ for a given nondegenerate $[b] = [b_1, b_2]$ and an unknown $[x] = [x_1, x_2]$. It follows that $b_1 + x_1 = 0, b_2 + x_2 = 0$ or

$$x_1 = -b_1, \ x_2 = -b_2. \tag{2}$$

However, since $[b] \in \mathbf{IR}$ we have $b_1 < b_2$ ($[b]$ is nondegenerate) and therefore $x_1 = -b_1 > -b_2 = x_2$. Hence the "solution" (2) does not define an element of $\mathbf{IR}$. In other words, if $b_1 \neq b_2$, the equation $[b] + [x] = 0$ has no solution in $\mathbf{IR}$. Similarly, if $b_1 \neq b_2$, then the equation $[b][x] = 1$ has no solution in $\mathbf{IR}$. Therefore in $\mathbf{IR}$, subtraction and division have to be introduced independently from addition and multiplication.

In addition, the distributive law does not hold in general. Instead we have the so-called *subdistributive law*

$$[a]([b] + [c]) \subseteq [a][b] + [a][c]. \tag{3}$$

We illustrate this by an example. Let $[a] = [2, 3]$, $[b] = [-2, -1]$, and $[c] = [1, 3]$. Then

$$[a]([b] + [c]) = [2, 3]([-2, -1] + [1, 3])$$
$$= [2, 3][-1, 2] = [-3, 6]$$

and

$$[a][b] + [a][c] = [2, 3][-2, -1] + [2, 3][1, 3]$$
$$= [-6, -2] + [2, 9] = [-4, 7].$$

Another property of the interval operations that is fundamental for all applications is the so-called *inclusion monotonicity*:

If $[a] \subseteq [c]$ and $[b] \subseteq [d]$ then $[a] * [b] \subseteq [c] * [d]$ for all $* \in \{+, -, \cdot, /\}$.

An (n, n) *interval matrix* is a square array $[A] = ([a]_{ij})$ whose entries $[a]_{ij}$ are elements of $\mathbf{IR}$. An *interval vector* $[a] = ([a]_i)$ is defined similarly. The set of interval matrices and the set of interval vectors are denoted by $M_n(\mathbf{IR})$ and $V_n(\mathbf{IR})$, respectively. Real matrices $A = (a_{ij}), a_{ij} \in \mathbf{R}$, are special elements of $M_n(\mathbf{IR})$. The analogue holds for real vectors $a = (a_i), a_i \in \mathbf{R}$.

In $V_n(\mathbf{IR})$ addition and subtraction are defined by

$$[a] \pm [b] = ([a]_i \pm [b]_i).$$

Similarly, in $M_n(\mathbf{IR})$ we define addition and subtraction of two elements by

$$[A] \pm [B] = ([a]_{ij} \pm [b]_{ij})$$

and multiplication by

$$[A][B] = \left(\sum_{j=1}^{n} [a]_{ij} [b]_{jk} \right).$$

The product of the interval matrix $[A]$ and the interval vector $[x]$ is

$$[A][x] = \left(\sum_{j=1}^{n} [a]_{ij} [x]_{ij} \right).$$

A set of laws that hold for these operations can be found in [2], Chapter 10.

The Gaussian Algorithm

Let $[A]$ be an interval matrix and $[b]$ an interval vector. We assume that the inverse A^{-1} exists for all $A \in [A]$. Then we define the *solution set* $\mathcal{S}$ (of $[A]$ and $[b]$) as

$$\mathcal{S} = \{x | Ax = b, A \in [A], b \in [b]\}.$$

$\mathcal{S}$ is a compact set. A more detailed characterization of the shape of $\mathcal{S}$ can be found in [4], where it is proved that in every orthant $\mathcal{S}$ is a convex polytope. See also [7]. Our goal is to enclose $\mathcal{S}$ by an interval vector $[x]$ that has this property: if $[A]$ and $[b]$ are shrinking to a real matrix A and a real vector b, respectively, then $[x]$ will shrink to $A^{-1}b$.

Starting with the so-called extended coefficient tableau

$$\begin{pmatrix} [a]_{11} & \cdots & [a]_{1n} & \Big| & [b]_1 \\ \vdots & \vdots & \vdots & \Big| & \vdots \\ [a]_{n1} & \cdots & [a]_{nn} & \Big| & [b]_n \end{pmatrix}$$

we form a new tableau

$$\begin{pmatrix} [a]'_{11} & [a]'_{12} & \cdots & [a]'_{1n} & \Big| & [b]'_1 \\ 0 & [a]'_{22} & \cdots & [a]'_{2n} & \Big| & [b]'_2 \\ \vdots & \vdots & & \vdots & \Big| & \vdots \\ 0 & [a]'_{n2} & \cdots & [a]'_{nn} & \Big| & [b]'_n \end{pmatrix}$$

by the formulae

$$\begin{cases} [a]'_{1j} = [a]_{1j} & j = 1, \ldots, n, \quad [b]'_1 = [b]_1 \\ [a]'_{ij} = [a]_{ij} - [a]_{1j} \left([a]_{i1} / [a]_{11} \right), & i, j = 2, \ldots n \\ [b]'_i = [b]_i - [b]_1 \left([a]_{i1} / [a]_{11} \right), & i = 2, \ldots, n \\ [a]'_{i1} = 0, & i = 2, \ldots, n. \end{cases} \tag{4}$$

provided $0 \notin [a]_{11}$.

We set $[A]' = ([a]'_{ij})$, $[b]' = ([b]'_i)$ and show that for the solution set

$$S' = \left\{ x \,|\, A'x = b', A' \in [A]', b' \in [b]' \right\}$$

we have $S \subseteq S'$. In order to prove this consider the real system $Ax = b$ where $A = (a_{ij}) \in [A]$, $b = (b_i) \in [b]$ are arbitrary but fixed. We are now eliminating the unknown x_1 from the second to the last equation and get a new system $A'x = b'$ which has the same solution as the original one. The elements of $A' = (a'_{ij})$ and $b' = (b'_i)$, respectively, are computed by the formulae of the first step of the Gaussian algorithm:

$$\begin{cases} a'_{1j} = a_{1j}, & j = 1, \ldots, n, \quad b'_1 = b_1 \\ a'_{ij} = a_{ij} - a_{1j}(a_{i1}/a_{11}), & i, j = 2, \ldots, n \\ b'_i = b_i - b_1(a_{i1}/a_{11}), & i = 2, \ldots, n \\ a_{i1} = a_{i1} - a_{11}(a_{i1}/a_{11}) = 0), & 1 = 2, \ldots, n. \end{cases} \tag{5}$$

Of course computation in the last line is not performed since the result $a'_{i1} = 0$ is known in advance. However, this line expresses the idea of the Gaussian elimination applied to systems with real coefficients: By using the existence of inverses with respect to addition and multiplication we are able to eliminate x_1 from row two to row n.

We cannot argue in the same manner when starting with the system with interval data. At first glance, setting $[a]'_{i1} = 0$, $i = 2, \ldots, n$, in (4) seems to make no sense. Recall that in general **IR** there exist no inverses with respect to addition and multiplication. Hence $[a]'_{i1} = 0$ cannot be achieved by an eliminating process. However, taking into account the inclusion monotonicity of the interval operations we have by (5) and (4) that $A' \in [A]', b' \in [b]'$ and therefore that $S \subseteq S'$.

Proceeding in this manner, the original coefficient tableau is transformed to the form

$$\left(\begin{array}{ccc|c} [\tilde{a}]_{11} & \cdots & [\tilde{a}] & [\tilde{b}]_1 \\ & \ddots & \vdots & \vdots \\ 0 & & [\tilde{a}]_{nn} & [\tilde{b}]_n \end{array} \right)$$

after $n - 1$ steps (provided all steps are feasible). Let $[\tilde{A}] = ([\tilde{a}]_{ij})$ and $[\tilde{b}] = ([\tilde{b}]_1)$. Then we have by induction

$$S \subseteq \tilde{S} = \left\{ x \,|\, \tilde{A}x = \tilde{b}, \; \tilde{A} \in [\tilde{A}], \; \tilde{b} \in [\tilde{b}] \right\}.$$

If $0 \notin [\tilde{a}]_{nn}$ then, using the formulae

$$[x]_n = [\tilde{b}]_n / [\tilde{a}]_{nn}$$

$$[x]_i = \left([\tilde{b}]_i - \sum_{j=i+1}^{n} [\tilde{a}]_{ij} [x]_j / [\tilde{a}]_{ii} \right), \quad i = n-1, \ldots, 1,$$

we obtain an interval vector $[x] = ([x]_i)$ satisfying

$$\mathcal{S} \subseteq \tilde{\mathcal{S}} \subseteq [x].$$

The property $\tilde{\mathcal{S}} \subseteq [x]$ follows again by the inclusion monotonicity.

Example 1.

Let $[A] = \begin{pmatrix} [2,4] & [-2,0] \\ [-1,0] & [2,4] \end{pmatrix}; [b] = \begin{pmatrix} [-2,2] \\ [-2,2] \end{pmatrix}$. See [3]. Then applying the Gaussian algorithm yields

$$[x]_1 = [-4,4], \; [x]_2 = [-3,3].$$

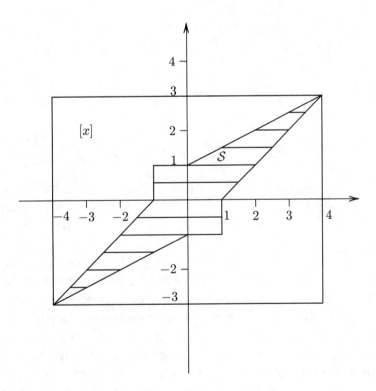

Figure 1. Optimality of the Gaussian algorithm

Figure 1 shows the solution set $\mathcal{S}$ (= the shaded region) and the interval vector $[x]$, which is in this case optimal, that is, there is no interval vector properly contained in $[x]$ and enclosing $\mathcal{S}$.

Example 2.

If we change the right-hand side to $[b] = \begin{pmatrix} [1,2] \\ [-2,2] \end{pmatrix}$ (see [3]), then the Gaussian algorithm delivers

$$[x]_1 = \left[-\frac{3}{2}, 4\right], \quad [x]_2 = [-2, 3]$$

which in this case is not an optimal enclosure of $\mathcal{S}$. See Figure 2. The solution set $\mathcal{S}$ can for this example already be found in [5].

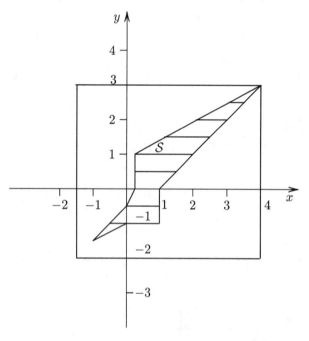

Figure 2. Nonoptimality of the Gaussian algorithm

Remarks

1) In applying the Gaussian algorithm to the given system with interval data, we assumed that all matrices $A \in [A]$ are nonsingular. As a consequence of this, there is at least one element $[a]_{i1}$ of the first column of $[A]$, which does not contain zero. Hence, after an eventually necessary exchange of rows the formulae (4) are applicable. However, it cannot be proved that this is also the case for the next steps. In other words, nonsingularity of all $A \in [A]$ does not guarantee the feasibility of the Gaussian algorithm, even if one allows row and/or column exchanges. This can be proved by counterexamples for the case $n \geq 3$. For $n = 2$, the feasibility is always guaranteed under the assumption mentioned. There exists a series

of sufficient criteria for feasibility. See [1] and [5], for example. However, in general this problem is unsolved.

2) An interesting question is how well $\mathcal{S}$ is included by the result $[x]$ of the Gaussian algorithm. In general there are smaller interval vectors $[y]$ than $[x]$ with the property $\mathcal{S} \subseteq [y]$. However, if $[A]$ is a so-called interval M-matrix and the right-hand side $[b]$ has a certain sign-distribution, then $[x]$ is the smallest interval vector for which $\mathcal{S} \subseteq [x]$. See Examples 1 and 2 and the discussion in [3].

3) More details and a complete overview on properties of the Gaussian algorithm known prior to 1992 can be found in the survey article [5].

4) We mention two simple applications of the Gaussian algorithm for linear systems with interval data:

a. If a real system has to be solved by the Gaussian algorithm, then usually the given data cannot be represented exactly on a computer with a given floating point system. If one encloses the data by the smallest representative intervals, then one has a linear system with interval data on the computer. Performing the Gaussian algorithm and taking into account the rounding errors, by rounding the exact bounds outwards to the nearest floating point numbers one computes an interval vector that contains the exact solution of the given real system.

b. If the solution of a nonlinear system is computed by the so-called interval Newton method, then the Gaussian algorithm for linear systems with interval data has to be performed in every step. For more details and additional applications see [2],[6], and [7].

References

1. G. Alefeld, Über die Durchführbarkeit des Gaußschen Algorithmus bei Gleichungen mit Intervallen als Koeffizienten. *Computing Suppl.* **1** (1977), 15–19.

2. G. Alefeld, J. Herzberger, *Introduction to Interval Computations*, Academic Press, New York, 1983.

3. W. Barth, E. Nuding, Optimale Lösung von Intervallgleichungssystemen, *Computing* **12** (1974), 117–125.

4. D.J. Hartfield, Concerning the Solution Set of $Ax = b$ where $P \leq A \leq Q$ and $p \leq b \leq q$, *Numer. Math.* **35** (1980), 355–359.

5. G. Mayer, Old and New Aspects for the Interval Gaussian Algorithm. In: *Computer Arithmetic, Scientific Computation and Mathematical Modeling*, E. Kaucher, S.M. Markov, G. Mayer (editors), J.C. Baltzer AG Scientific Publishing Co., Basel, 1992.

6. R.E. Moore, *Interval Analysis*, Prentice Hall, Englewood Cliffs, N.J., 1966.

7. A. Neumaier, *Interval Methods for Systems of Equations*, Cambridge University Press, Cambridge, 1990.

G. Alefeld and G. Mayer
Universität Karlsruhe
Contributed

On Sylvester's Law of Nullity

When one extends Sylvester's Law of Nullity* from square matrices over a field to rectangular matrices over a field, the question arises which are the most general conditions. The answer is as follows.

Sylvester's Law of Nullity. *The nullity of AB is less than or equal to the sum of the nullities of A and B. It is greater than or equal to the nullity of A; it is greater than or equal to that of B provided A does not have more columns than it has rows.*

Proof. Let $R(M)$ and $N(M)$ denote the rank and the nullity of the matrix M, and let A and B be $m \times n$ and $n \times p$ matrices respectively.

(a) Following standard procedure [1], choose non-singular $m \times m$ and $n \times n$ matrices P and Q such that the $m \times n$ matrix $PAQ = A^*$ is diagonal, having the entries of an identity matrix of rank $r = R(A)$ in the left upper corner and zeros elsewhere. Set $B^* = Q^{-1}B$, then $A^*, B^*, A^*B^* = PAB$ are equivalent to and therefore have the ranks and nullities of A, B, and AB, respectively. The first rows of A^*B^* are those of B^*, and the remaining $m - r$ rows consist of zeros; hence A^*B^*, which lacks $n - r$ possible independent rows of B^*, has a rank of at least $R(B^*) - (n - R(A^*))$. Therefore $N(A^*B^*) = m - R(A^*B^*) \leq m - R(A^*) + n - R(B^*) = N(A^*) + N(B^*)$.

(b) The null space of A is clearly a subspace of that of AB, hence $N(A) \leq N(AB)$.

(c) Since $R(AB) \leq R(B)$, we have $N(B) \leq m - R(B) \leq N(AB)$ provided $n \leq m$. We may have $N(AB) < N(B)$ if $n > m$, as is shown by the example

$$A = \begin{bmatrix} 1 & 0 \end{bmatrix}, \qquad B = \begin{bmatrix} 1 & 0 \\ 0 & 0 \end{bmatrix}, \qquad AB = A, \qquad N(AB) = 0, \qquad N(B) = 1.$$

Reference

1. G. Birkoff and S. MacLane, *A Survey of Modern Algebra*, revised edition, New York, 1953, p. 235.

Kurt Bing
Rensselaer Polytechnic Institute
American Mathematical Monthly **64** (1957), 100

Editors' Note: Sylvester's Law of Nullity: Let $T : U \to V$ and $S : V \to W$ be linear mappings of finite-dimensional vector spaces. Then nullity$(ST) \leq$ nullity$(T) +$ nullity(S).

Inverses of Vandermonde Matrices

It is frequently useful to be able to produce the inverse of a Vandermonde matrix for curve fitting, numerical differentiation, and difference equations. Explicit formulas have been given in [1], and this note simplifies those formulas to the point where pencil and paper calculation for the inverse of a matrix of order six takes about twenty minutes.

Let $V(n)$ be the Vandermonde matrix

$$V(n) = \begin{bmatrix} 1 & 1 & \cdots & 1 \\ x_1 & x_2 & \cdots & x_n \\ x_1^2 & x_2^2 & \cdots & x_n^2 \\ \cdots\cdots\cdots\cdots\cdots\cdots\cdots \\ x_1^{n-1} & x_2^{n-1} & \cdots & x_n^{n-1} \end{bmatrix}$$

We define $F(x)$ to be a polynomial whose roots are $x_i, i = 1, 2, \ldots, n$, and $f_k(x)$ to be the reduced polynomial when the factor $x_k - x$ is taken from $F(x)$. Thus

$$F(x) = \prod_{i=1}^{n}(x_i - x), \quad \text{and} \quad f_k(x) = F(x)/(x_k - x).$$

A useful result follows immediately, namely that

(1) $$f_k(x_j) = 0 \text{ for } j \neq k, \quad \text{and} \quad f_k(x_k) = -F'(x_k)$$

where $'$ indicates the first derivative. The first part of this result follows from the definition of $f_k(x)$ (only the factor $(x_k - x)$ is taken from $F(x)$). $F'(x)$ will be the sum of n terms, each one of which contains $n - 1$ factors, and each term except one contains the factor $x_k - x$. When $x = x_k$, this nonzero term is precisely $-f_k(x_k)$, which establishes the second part.

To form the inverse, write the co-efficients of $-f_k(x)/F'(x_k)$, in row k, $k = 1, 2, \ldots, n$ to form the matrix M. We see that the element in row k, column j of MV is precisely $-f_k(x_j)/F'(x_k)$, and result (1) establishes that M is the required inverse.

In actual practice, we can calculate $F(x)$, then divide synthetically by $x_1 - x$ to find a vector orthogonal to each of the column vectors of $V(n)$ except the first. This vector can be normalized by dividing each element by $-f_1(x_1)$. The remaining rows can be quickly calculated in a similar manner.

Reference

1. N. Macon and A. Spitzbart, Inverses of Vandermonde Matrices, *American Mathematical Monthly* **65** (1958), 95–100.

F.D. Parker
University of Alaska
American Mathematical Monthly **71** (1964), 410–411

On One-sided Inverses of Matrices

Standard texts in linear algebra discuss at some length the notion of invertibility of a matrix with n rows and n columns. Very little, if anything, is usually said about one-sided inverses of a non-square matrix. The purpose of this article is to provide complete answers to the following questions:

(a) What are necessary and sufficient conditions for a matrix to possess a left (right) inverse?

(b) How can we compute all left (right) inverses of a matrix once we know that they exist?

Our discussion uses for a point of departure several standard definitions and theorems from linear algebra. We describe some of them in the next paragraph. No proofs are given, since the reader should be able to find them in most introductory linear algebra texts.

The *rank* of matrix $A = [a_{ij}]_{m,n}$ over a field of scalars F, symbolically $r(A)$, is the dimension of the vector space spanned by its column vectors. It is a theorem that the rank of A equals the dimension of the vector space spanned by its row vectors (*row rank*). A matrix $A = [a_{ij}]_{n,n}$ is *invertible* if and only if there exists a matrix $B = [b_{ij}]_{n,n}$ such that $AB = BA = I_n$, where I_n is the identity matrix with n rows and n columns. The *Kronecker delta* δ_{ij} is defined as follows: $\delta_{ij} = 1$ if $i = j$ and $\delta_{ij} = 0$ if $i \neq j$. If the rank of $A = [a_{ij}]_{m,n}$ is k, then there exist invertible matrices P and Q such that $PAQ = [c_{ij}]_{m,n}$, where $c_{ij} = \delta_{ij}$ if $j \leq k$ and $c_{ij} = 0$ otherwise. The matrix $[c_{ij}]_{m,n}$ described in the preceding sentence is called the *normal form* of A. A matrix B is a *left inverse* of $A = [a_{ij}]_{m,n}$ if and only if B is a solution of the matrix equation $XA = I_n$. Similarly, a *right inverse* of $A = [a_{ij}]_{m,n}$ is a solution of the equation $AY = I_m$. We use the symbol $\hat{A}_\ell$ ($\hat{A}_r$) to denote the set of all left (right) inverses of A. If a matrix A is invertible, then for any matrices B and C $r(AB) = r(B)$ and $r(CA) = r(C)$, whenever the compositions AB, CA make sense.

Next we address ourselves to the solution of our problem. It will be understood that all matrices under consideration have their entries selected from a fixed field of scalars F.

Theorem 1. Consider the matrix $A = [a_{ij}]_{m,n}$ with m rows and n columns. Then

(a) A has a left inverse if and only if $r(A) = n$,

(b) A has a right inverse if and only if $r(A) = m$.

209

Proof.

(a) Suppose $r(A) = n$. We assert that for $i = 1, 2, \ldots, n$ there exist scalars $x_{i1}, x_{i2}, \ldots, x_{im}$ such that

$$x_{i1} \cdot \begin{bmatrix} a_{11} \\ a_{12} \\ \vdots \\ a_{1n} \end{bmatrix} + x_{i2} \cdot \begin{bmatrix} a_{21} \\ a_{22} \\ \vdots \\ a_{2n} \end{bmatrix}$$

$$+ \ldots + x_{im} \cdot \begin{bmatrix} a_{m1} \\ a_{m2} \\ \vdots \\ a_{mn} \end{bmatrix} = \begin{bmatrix} \delta_{i1} \\ \delta_{i2} \\ \vdots \\ \delta_{in} \end{bmatrix}$$

The truth of our assertion becomes apparent if we note that the row rank of $A = n \leq m$ and observe that the vectors

$$\begin{bmatrix} a_{j1} \\ a_{j2} \\ \vdots \\ a_{jn} \end{bmatrix} \qquad j = 1, 2, \ldots, m$$

are just the row vectors of A written in column form. Now let $X = [x_{ij}]_{n,m}$. Then $XA = [\delta_{ij}]_{n,m} = I_n$ because of the way each x_{ij} was selected. To prove the converse, suppose there is a matrix $X = [x_{ij}]_{n,m}$ such that $XA = I_n$. Let $A_1, A_2, \ldots, A_n$ denote the column vectors of A and suppose there are scalars $c_1, c_2, \ldots, c_n$ such that $\sum_{j=1}^{n} c_j A_j = 0$. Since X is a left inverse of A, its ith row vector X_i has the property $X_i A_j = 0$, if $i \neq j$ and $X_i A_i = 1$. Multiplying the equation $\sum_{j=1}^{n} c_j A_j = 0$ successively by $X_1, X_2, \ldots, X_n$ we find that $c_1 = c_2 = \cdots = c_n = 0$. Hence the set $\{A_1, A_2, \ldots, A_n\}$ is linearly independent and $r(A) = n$.

(b) Recall that for a matrix $A = [a_{ij}]_{m,n}$ it *transpose* $A^t = [a_{ij}^*]_{n,m}$ where $a_{ij}^* = a_{ji}$ for each i and j. It is a theorem that $(AB)^t = B^t A^t$ whenever the composition AB makes sense. Using part (a) we reason that A has a right inverse B implies $AB = I_m$. Hence $B^t A^t = I_m$. From part (a), $m = r(A^t) = r(A)$. Conversely, if $r(A) = m$, then $r(A^t) = m$. Consequently there is a matrix D such that $DA^t = I_m$. Taking transposes, we obtain $AD^t = I_m$ which shows that A has a right inverse.

Some easy consequences of this theorem, applied to the matrix $A = [a_{ij}]_{m,n}$, are:

(a) If $r(A) < m$, then A has no right inverse.
(b) If $r(A) < n$, then A has no left inverse.
(c) If $n < m$, then A has no right inverse.
(d) If $m < n$, then A has no left inverse.

(e) If $r(A) < n \leq m$ or $r(A) < m \leq n$, then A has neither a right nor a left inverse.

Let us now attack the problem of finding all left inverses of $A = [a_{ij}]_{m,n}$, given that $r(A) = n$.

Theorem 2. Suppose $A = [\delta_{ij}]_{m,n}$ and $r(A) = n$ (that is, A is in normal form). Then $B = [b_{ij}]_{n,m}$ is a left inverse of A if and only if $b_{ij} = \delta_{ij}$ for $i, j \epsilon \{1, 2, \ldots, n\}$.

Proof. From $r(A) = n$ it follows that $n \leq m$. Let $BA = [c_{ij}]_{n,n}$ with $c_{ij} = \sum_{k=1}^{m} b_{ik}\delta_{ij}$. Since $I_n = [\delta_{ij}]_{n,n}$, we conclude that $BA = I_n$ if and only if $b_{ij} = \delta_{ij}$ for $i, j \epsilon \{1, 2, \ldots, n\}$.

Example.

Let $A = \begin{bmatrix} 1 & 0 \\ 0 & 1 \\ 0 & 0 \end{bmatrix}$. According to Theorem 2, $B \in \hat{A}_e$ if and only if $B = \begin{bmatrix} 1 & 0 & x \\ 0 & 1 & y \end{bmatrix}$ for arbitrary scalars x and y. Hence we have solved the matrix equation $XA = I_2$.

Theorem 3. Let $A = [a_{ij}]_{m,n}$ be a matrix with $r(A) = n$. Let P and Q be invertible matrices such that $PAQ = [\delta_{ij}]_{m,n}$. Then $B \in \hat{A}_\ell$ if and only if $B = QDP$ where D is a left inverse of $[\delta_{ij}]_{m,n}$.

Proof. If $B \in \hat{A}_\ell$, then $Q^{-1}BP^{-1}[\delta_{ij}]_{m,n} = Q^{-1}BP^{-1}(PAQ) = I_n$. We have shown now that $D = Q^{-1}BP^{-1}$ is a left inverse of $[\delta_{ij}]_{m,n}$. It follows that $B = QDP$. On the other hand, if D is a left inverse of $[\delta_{ij}]_{m,n}$ and $B = QDP$, then

$$BA = B\left(P^{-1}[\delta_{ij}]_{m,n}Q^{-1}\right)$$
$$= (QDP)\left(P^{-1}[\delta_{ij}]_{m,n}Q^{-1}\right) = I_n.$$

The significance of Theorem 3 becomes apparent if we consider that, according to our introductory remarks, if $r(A) = n$, invertible matrices P and Q do exist such that $PAQ = [\delta_{ij}]_{m,n}$. Thus Theorem 3 in conjunction with Theorem 2 provides us with a computational tool for solving the matrix equation $XA = I_n$, provided we can find invertible matrices P and Q such that $PAQ = [\delta_{ij}]_{m,n}$. However, the latter task is well known and a computational procedure is given in most texts on linear algebra.

Example.

Let us solve the matrix equation

$$XA = I_3 \text{ where } A = \begin{bmatrix} 1 & 0 & 1 \\ 2 & 1 & 1 \\ 3 & 0 & 1 \\ 0 & 1 & 1 \end{bmatrix}.$$

We find that $r(A) = 3$ and conclude that the given equation has solutions. Next we find invertible matrices P and Q such that $PAQ = [\delta_{ij}]_{4,3}$. The technique is displayed as follows.

$$
\left[\begin{array}{cccc|ccc}
1 & 0 & 0 & 0 & 1 & 0 & 1 \\
0 & 1 & 0 & 0 & 2 & 1 & 1 \\
0 & 0 & 1 & 0 & 3 & 0 & 1 \\
0 & 0 & 0 & 1 & 0 & 1 & 1
\end{array}\right]
$$

$$
\begin{array}{c}
-3R_1 + R_3 \\
-2R_1 + R_2 \\
\xrightarrow{\hspace{2cm}}
\end{array}
$$

$$
\left[\begin{array}{cccc|ccc}
1 & 0 & 0 & 0 & 1 & 0 & 1 \\
-2 & 1 & 0 & 0 & 0 & 1 & -1 \\
-3 & 0 & 1 & 0 & 0 & 0 & -2 \\
0 & 0 & 0 & 1 & 0 & 1 & 1
\end{array}\right]
$$

$$
\begin{array}{c}
-R_2 + R_4 \\
\xrightarrow{\hspace{2cm}}
\end{array}
$$

$$
\left[\begin{array}{cccc|ccc}
1 & 0 & 0 & 0 & 1 & 0 & 1 \\
-2 & 1 & 0 & 0 & 0 & 1 & -1 \\
-3 & 0 & 1 & 0 & 0 & 0 & -2 \\
2 & -1 & 0 & 1 & 0 & 0 & 2
\end{array}\right]
$$

$$
\begin{array}{c}
-1/2R_3 \\
\xrightarrow{\hspace{2cm}}
\end{array}
$$

$$
\left[\begin{array}{cccc|ccc}
1 & 0 & 0 & 0 & 1 & 0 & 1 \\
-2 & 1 & 0 & 0 & 0 & 1 & -1 \\
3/2 & 0 & 1/2 & 0 & 0 & 0 & 1 \\
2 & -1 & 0 & 1 & 0 & 0 & 2
\end{array}\right]
$$

$$
\begin{array}{c}
-R_3 + R_1 \\
R_3 + R_2 \\
-2R_3 + R_4 \\
\xrightarrow{\hspace{2cm}}
\end{array}
$$

$$
\left[\begin{array}{cccc|ccc}
-1/2 & 0 & 1/2 & 0 & 1 & 0 & 0 \\
-1/2 & 1 & -1/2 & 0 & 0 & 1 & 0 \\
3/2 & 0 & -1/2 & 0 & 0 & 0 & 1 \\
-1 & -1 & 1 & 1 & 0 & 0 & 0
\end{array}\right]
$$

Thus we can choose

$$P = \begin{bmatrix} -1/2 & 0 & 1/2 & 0 \\ -1/2 & 1 & -1/2 & 0 \\ 3/2 & 0 & -1/2 & 0 \\ -1 & -1 & 1 & 1 \end{bmatrix} \text{ and } Q = I_3.$$

The reader should convince himself that in general if $A = [a_{ij}]_{m,n}$ and $r(A) = n$, then there exists an invertible matrix P such that $PA = [\delta_{ij}]_{m,n}$. According to Theorem 2, every left inverse of $[\delta_{ij}]_{4,3}$ has the form

$$\begin{bmatrix} 1 & 0 & 0 & x_1 \\ 0 & 1 & 0 & x_2 \\ 0 & 0 & 1 & x_3 \end{bmatrix}.$$

Using Theorem 3 we conclude that $B \in \hat{A}_\ell$ if and only if

$$B = \begin{bmatrix} 1 & 0 & 0 & x_1 \\ 0 & 1 & 0 & x_2 \\ 0 & 0 & 1 & x_3 \end{bmatrix} \cdot P, \text{ that is}$$

$$B = \begin{bmatrix} -1/2 - x_1 & -x_1 & 1/2 + x_1 & x_1 \\ -1/2 - x_2 & 1 - x_2 & -1/2 + x_2 & x_2 \\ 3/2 - x_3 & -x_3 & -1/2 + x_3 & x_3 \end{bmatrix}$$

where x_1, x_2, x_3 are any scalars.

This ends our discussion on left inverses of a matrix. The important facts on right inverses are stated in our next theorem.

Theorem 4. Let $A = [a_{ij}]_{m,n}$ with $r(A) = m$. Suppose P and Q are invertible matrices such that $PAQ = [\delta_{ij}]_{m,n}$. Then $B \in \hat{A}_r$ if and only if $B = QDP$ where D is a right inverse of $[\delta_{ij}]_{m,n}$. A matrix C is a right inverse of $[\delta_{ij}]_{m,n}$ if and only if C^t is a left inverse of $[\delta_{ij}]_{n,m}$.

Proof. Let $B \in \hat{A}_r$. $[\delta_{ij}]_{m,n} \cdot Q^{-1}BP^{-1} = (PAQ)(Q^{-1}BP^{-1}) = I_m$. Hence $D = Q^{-1}BP^{-1}$ is a right inverse of $[\delta_{ij}]_{m,n}$. It follows that $B = QDP$. Conversely, if $B = QDP$ with $[\delta_{ij}]_{m,n} \cdot D = I_m$, then $AB = (P^{-1}[\delta_{ij}]_{m,n}Q^{-1})(QDP) = I_m$. The last statement of the theorem follows readily if we observe that $[\delta_{ij}]_{m,n}^t = [\delta_{ij}]_{n,m}$.

The reader should observe that Theorem 4 in conjunction with Theorem 2 provides us with means to solve the matrix equation $AX = I_m$. An instructor searching for a worthwhile project in linear algebra could ask his students to flow-chart a procedure for finding left or right inverses of matrices. A successfully run computer program should enhance the student's interest in linear algebra while requiring him to display his understanding of the underlying theory.

Elmar Zemgalis
Highline Community College
College Mathematics Journal **2** (1971), 45–48

Integer Matrices Whose Inverses Contain Only Integers

If a square matrix and its inverse contain only integers, the matrix will be called *nice*. A simple method for constructing nice matrices will be given and some of the uses of nice matrices will be discussed. Then a proof of the validity of the method will be given. Finally it will be shown that this method does in fact generate *all* nice matrices

The following method shows how to construct nice matrices.

(1) Form a triangular integer matrix A with all zero entries below (or above) the main diagonal, with elements on the main diagonal chosen so their product is ± 1; for example,

$$A = \begin{bmatrix} 1 & -2 & 3 \\ 0 & -1 & 4 \\ 0 & 0 & 1 \end{bmatrix}.$$

(2) Let θ be an elementary row or column operation other than multiplying any row or column by a constant $\neq \pm 1$. Any such operation may be applied to the initial matrix and may be followed by a similar operation as often as desired.

Example 1. Let θ_1: multiply row 1 by row 2 and add to row 2. Then

$$\theta_1(A) = \begin{bmatrix} 1 & -1 & 3 \\ 2 & -5 & 10 \\ 0 & 0 & 1 \end{bmatrix} = B.$$

Let θ_2: multiply row 1 by -1 and add to row 3. then

$$\theta_2(B) = \begin{bmatrix} 1 & -2 & 3 \\ 2 & -5 & 10 \\ -1 & 2 & -2 \end{bmatrix} = C, \text{ etc.}$$

Matrices A, B, and C are all nice matrices.

Example 2. Consider

$$A = \begin{bmatrix} 1 & 0 & 0 & 0 \\ 3 & -1 & 0 & 0 \\ 0 & 1 & -1 & 0 \\ 2 & 4 & 3 & 1 \end{bmatrix}.$$

215

Let θ_1: add 2 times row 4 to row 1 to get

$$\theta_1(A) = \begin{bmatrix} 5 & 8 & 6 & 2 \\ 3 & -1 & 0 & 0 \\ 0 & 1 & -1 & 0 \\ 2 & 4 & 3 & 1 \end{bmatrix} = B.$$

Then let θ_2: add -3 times column 1 to column 4 to get

$$\theta_2(B) = \begin{bmatrix} 5 & 8 & 6 & -13 \\ 3 & -1 & 0 & -9 \\ 0 & 1 & -1 & 0 \\ 2 & 4 & 3 & -5 \end{bmatrix} = C.$$

This method will provide many examples of nice matrices quickly and easily.

As students quickly learn, most matrices are not invertible, let alone nice. Thus an easily constructible supply of nice matrices can be most useful for students and teachers. For example, if the coefficients for the variables in a system of linear equations with integer constants form a nice matrix, the solutions to the system will be integers and easy to check. Moreover, the arithmetic of the solution, whether found by elimination and substitution, row reduction of the augmented matrix, or matrix inverse methods, will involve mostly integers. Thus the student can concentrate more on the technique being learned with less chance for computational errors.

In learning to calculate the inverse of a nonsingular matrix, nice matrices are excellent tools. The inverse will be computationally easy to find and easy to check. Students can also generate their own nice matrices and find the corresponding inverses to develop individual message coding-decoding systems as in the following example.

Individual letters, words, and symbols are assigned arbitrary integer values; for example,

$$A = 1, \quad B = -1, \quad C = 2, \quad D = -2, \dots, \quad Z = -13, \quad \text{space} = 0.$$

A nice matrix A is chosen as an encoder, and its inverse A^{-1} will be the decoder, say

$$A = \begin{bmatrix} 1 & -2 & 3 \\ 2 & -5 & 10 \\ -1 & 2 & -2 \end{bmatrix} \quad A^{-1} = \begin{bmatrix} 10 & -2 & 5 \\ 6 & -1 & 4 \\ 1 & 0 & 1 \end{bmatrix}.$$

A message like "COME HOME" would be represented first as

$$2, 8, 7, 3, 0, -4, 8, 7, 3.$$

The message would then be encoded by premultiplying each set of three integers by matrix A.

$$\begin{bmatrix} 1 & -2 & 3 \\ 2 & -5 & 10 \\ -1 & 2 & -2 \end{bmatrix} \cdot \begin{bmatrix} 2 \\ 8 \\ 7 \end{bmatrix}, \begin{bmatrix} 3 \\ 0 \\ -4 \end{bmatrix}, \begin{bmatrix} 8 \\ 7 \\ 3 \end{bmatrix} = \begin{bmatrix} 7 \\ 34 \\ 0 \end{bmatrix}, \begin{bmatrix} -9 \\ -34 \\ 5 \end{bmatrix}, \begin{bmatrix} 3 \\ 11 \\ 0 \end{bmatrix}.$$

The coded message would be

$$7, 34, 0, 15, -34, -5, 3, 11, 0.$$

To decode the message each set of three coded numbers would be premultiplied by A^{-1} and checked against the initial integer assignment. Here

$$\begin{bmatrix} 10 & -2 & 5 \\ 6 & -1 & 4 \\ 1 & 0 & 1 \end{bmatrix} \cdot \begin{bmatrix} 7 \\ 34 \\ 0 \end{bmatrix} = \begin{bmatrix} 2 \\ 8 \\ 7 \end{bmatrix}, \text{ etc.}$$

Here is the reason this method for generating nice matrices works:

$$A^{-1} = \frac{1}{\text{Det}A}(\text{Adj}A),$$

where $\text{Adj}A$ is the transpose of the matrix obtained by replacing each element a_{ij} of A by its cofactor c_{ij}. Since $c_{ij} = (-1)^{i+j}$ times the determinant of the submatrix of A obtained by deleting row i and column j, each c_{ij} is an integer if A contains only integers. Hence $\text{Adj}A$ will contain only integers if A does; and A will contain all integers if $1/(\text{Det}A)$ is $+1$ or -1. Now the determinant of a triangular matrix equals the product of the elements on the main diagonal. Thus the determinant of the original matrix chosen will equal $+1$ or -1. It is also true that any elementary row operation preserves the determinant. Hence A will consist of only integer entries.

A proof that all nice matrices are generated by this method will now be given. Let B be an arbitrary nice matrix of dimensions $n \times n$. For matrices P and Q define $P \sim Q$ if there exists a finite sequence $\theta_1, \theta_2, \ldots, \theta_t$ of operations θ given in (2) above such that $\theta_t(\cdots(\theta_2(\theta_1(P)))\cdots) = Q$. It is easy to see that $\sim$ is an equivalence relation. It will be shown that there is a triangular matrix A of initial form (1) above such that $B \sim A$. Thus $A \sim B$ and the proof will be complete.

The first column of B must contain some nonzero integers else $\text{Det}B = 0$ and B would be singular. Let row i contain a nonzero integer of least absolute value in column 1. By adding proper multiples of row i to the other rows, the absolute values of the integers in column 1 will be reduced. This process can be continued using other rows till after a finite number of steps only one nonzero integer remains in column 1. After a row interchange one has

$$B \sim \begin{bmatrix} k_1 & * & \cdots & * \\ 0 & * & \cdots & * \\ \vdots & & & \\ 0 & * & \cdots & * \end{bmatrix}.$$

Applying the same technique to successively smaller submatrices leads finally to

$$B \sim \begin{bmatrix} k_1 & * & * & \cdots & * \\ 0 & k_2 & * & \cdots & * \\ 0 & 0 & k_3 & \cdots & * \\ \cdot & \cdot & \cdot & & \\ 0 & 0 & 0 & \cdots & k_n \end{bmatrix} = A.$$

Since B and B^{-1} are integer matrices and since $\mathrm{Det}(B \cdot B^{-1}) = \mathrm{Det}B \cdot \mathrm{Det}B^{-1} = \mathrm{Det}I = 1$, it follows that $\mathrm{Det}B = \pm 1$. Now if $P \sim Q$ then $\mathrm{Det}P = \mathrm{Det}Q$. Thus $\mathrm{Det}A = \mathrm{Det}B$ or $k_1 k_2 \cdots k_n = \pm 1$, and the matrix A has the desired initial form.

In closing we note that

(1) the initial generating matrix for a given nice matrix is clearly not unique, and

(2) all nice matrices could actually be generated from diagonal matrices with ± 1 on the main diagonal. Thus the set of all nice matrices of dimension $n \times n$ is the equivalence class of all matrices equivalent to I_n under the equivalence relation $\sim$ indicated above.

Robert Hanson
James Madison University
College Mathematics Journal **13** (1982), 18–21

PART 8
Applications

Introduction

Linear algebra has applications in almost every area of mathematics as well as in numerous disciplines outside of mathematics. Many textbooks now contain a variety of applications to other fields such as biology, chemistry, computer science, engineering, operations research, physics, and statistics. Within mathematics, linear algebra is used in fields such as analysis, combinatorics, control theory, geometry, linear programming, and numerical analysis. The papers included in this part describe applications within mathematics.

The Matrix-Tree Theorem

When one examines the standard textbooks for a first course in linear algebra (such as [1],[6]), one finds an introductory chapter dealing with the concept and properties of determinants. Then the determinant is usually tied to the major theme of the matrix through the relationship: An $n \times n$ matrix A is invertible if and only if the determinant of A is not zero. Often Cramer's Rule is then introduced as a method for solving a system of n linear (independent) equations in n unknowns via determinants. And so the student finds himself, or herself, evaluating some 2×2, 3×3, or maybe even 4×4 determinants.

Larger determinants seem to be avoided. After all, where do linear systems (larger than three equations in three unknowns) come about in realistic situations? Depending upon his or her background, the student may feel that such systems only arise as textbook exercises—rather tedious efforts in computational boredom, at best. However, recent changes in curriculum and technology show why and how larger determinants should be examined. In particular,

(1) Structures from discrete mathematics can be naturally brought into the introductory algebra course and provide some ideas that involve determinants.

(2) The use of a computer algebra system, such as MAPLE, can drastically reduce the time spent on the drudgery of computing large determinants.

In order to accomplish our goal we shall use some ideas from graph theory. We shall introduce some of the basic ideas and refer the reader to [4] for additional reading (if necessary).

The Matrix-Tree Theorem

The Matrix-Tree Theorem is not a new result—in fact, it has been around for well over fifty years. It was apparently first discovered in 1940 by Brooks, Smith, Stone, and Tutte from results in the study of electrical networks—some results going back to the 1847 paper of Kirchhoff. The theorem tells us how to determine the number of spanning trees for an undirected graph by means of a determinant. Those interested in a formal proof of the result should consult [2] or [5].

An undirected graph $G = (V, E)$ consists of a nonempty set V of vertices and a set E of unordered pairs $\{x, y\} (= \{y, x\})$, called edges, where $x, y \in V$ and $x \neq y$. In Figure 1(a) we have the undirected graph $G = (V, E)$ where $V = \{1, 2, 3, 4, 5\}$ and $E = \{\{1, 2\}, \{1, 3\}, \{1, 5\}, \{2, 3\}, \{3, 4\}, \{4, 5\}\}$. Figures 1(b) and (c) provide two connected subgraphs of G, where all of the vertices in V are used, and where no cycle of edges appears. In G, the edge sets $\{\{1, 2\}, \{2, 3\}, \{3, 1\}\}$ and $\{\{1, 3\}, \{3, 4\}, \{4, 5\}, \{5, 1\}\}$ provide examples of cycles. The subgraphs in Figures 1(b) and (c) are examples of spanning trees of G.

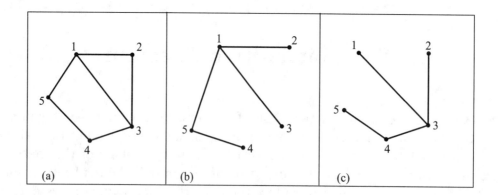

Figure 1

Note: These spanning trees are nonidentical because they gave different edge sets. However, they are isomorphic: Correspond vertex 1 in (b) with vertex 3 in (c), 2 with 2, 3 with 1, 4 with 5, and 5 with 4.

Our objective here is to count all of the nonidentical spanning trees of G. In order to do so we represent the graph G by means of its adjacency matrix $A(G)$. Here $A(G)$ is a symmetric $5 \times 5(0, 1)$-matrix, where the rows and columns are indexed by the set V—fixed as 1,2,3,4,5. That is,

$$A(G) = (a_{ij})_{5 \times 5}, \text{ where } a_{ij} = 1 \text{ if } \{i, j\} \text{ is an edge in } G, \text{ and } a_{ij} = 0, \text{ otherwise.}$$

Consequently,

$$A(G) = \begin{array}{c} \\ (1) \\ (2) \\ (3) \\ (4) \\ (5) \end{array} \begin{array}{ccccc} (1) & (2) & (3) & (4) & (5) \\ \left[\begin{array}{ccccc} 0 & 1 & 1 & 0 & 1 \\ 1 & 0 & 1 & 0 & 0 \\ 1 & 1 & 0 & 1 & 0 \\ 0 & 0 & 1 & 0 & 1 \\ 1 & 0 & 0 & 1 & 0 \end{array}\right] \end{array} \quad \text{and} \quad D(G) = \begin{array}{c} \\ (1) \\ (2) \\ (3) \\ (4) \\ (5) \end{array} \begin{array}{ccccc} (1) & (2) & (3) & (4) & (5) \\ \left[\begin{array}{ccccc} 3 & 0 & 0 & 0 & 0 \\ 0 & 2 & 0 & 0 & 0 \\ 0 & 0 & 3 & 0 & 0 \\ 0 & 0 & 0 & 2 & 0 \\ 0 & 0 & 0 & 0 & 2 \end{array}\right] \end{array}$$

where $D(G) = (d_{ij})_{5 \times 5}$ is called the degree matrix of G. For $D(G)$ we have

$$d_{ij} = \begin{cases} \text{the number of edges that contain vertex } i, & \text{when } i = j \\ 0, & \text{otherwise.} \end{cases}$$

Then from the Matrix-Tree Theorem one learns that the number of nonidentical spanning trees for any connected labeled undirected graph G, where $|E| \geq 1$, is given by the value of *any* cofactor of the matrix $D(G) - A(G)$. Here

$$D(G) - A(G) = \begin{bmatrix} 3 & -1 & -1 & 0 & -1 \\ -1 & 2 & -1 & 0 & 0 \\ -1 & -1 & 3 & -1 & 0 \\ 0 & 0 & -1 & 2 & -1 \\ -1 & 0 & 0 & -1 & 2 \end{bmatrix}$$

and if we use the $(3,1)$-cofactor, then the theorem tells us that G has

$$(-1)^{3+1} \begin{vmatrix} -1 & -1 & 0 & -1 \\ 2 & -1 & 0 & 0 \\ 0 & -1 & 2 & -1 \\ 0 & 0 & -1 & 2 \end{vmatrix} \qquad \text{spanning trees.}$$

Using the computer algebra system MAPLE, the code given in Figure 2 evaluates (the determinant of) the cofactor above as 11 and tells us that the graph in Figure 1(a) has 11 nonidentical spanning trees.

```
> q1 := matrix(5, 5, [3, -1, -1, 0, -1, -1, 2, -1, 0, 0, -1, -1, 3, -1, 0, 0, 0, -1, 2, -1,
   -1, 0, 0, -1, 2]) :
> (-1)^(3 + 1) * det(minor(q1, 3, 1));
                                    11
```

Figure 2

Now the reader may feel that the preceding example could have been solved just as readily (if not, more easily) by simply drawing all of the 11 nonidentical spanning trees. And then there would be no need to consider determinants, cofactors, or any of the other mathematical ideas we mentioned. In order to put such skepticism in its place, consider the graph shown in Figure 3(a). This graph arises throughout the study of graph theory and is called the Petersen graph. In Figure 3(b) we have the matrix $D(G) - A(G)$ for this graph.

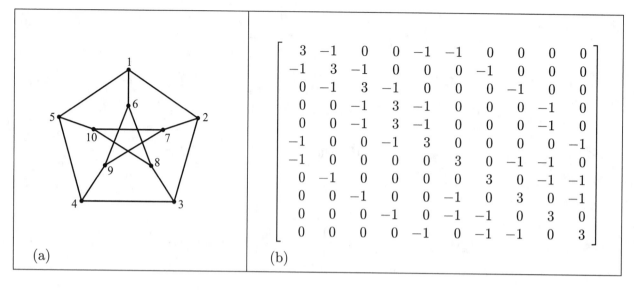

$$
\begin{bmatrix}
3 & -1 & 0 & 0 & -1 & -1 & 0 & 0 & 0 & 0 \\
-1 & 3 & -1 & 0 & 0 & 0 & -1 & 0 & 0 & 0 \\
0 & -1 & 3 & -1 & 0 & 0 & 0 & -1 & 0 & 0 \\
0 & 0 & -1 & 3 & -1 & 0 & 0 & 0 & -1 & 0 \\
0 & 0 & -1 & 3 & -1 & 0 & 0 & 0 & -1 & 0 \\
-1 & 0 & 0 & -1 & 3 & 0 & 0 & 0 & 0 & -1 \\
-1 & 0 & 0 & 0 & 0 & 3 & 0 & -1 & -1 & 0 \\
0 & -1 & 0 & 0 & 0 & 0 & 3 & 0 & -1 & -1 \\
0 & 0 & -1 & 0 & 0 & -1 & 0 & 3 & 0 & -1 \\
0 & 0 & 0 & -1 & 0 & -1 & -1 & 0 & 3 & 0 \\
0 & 0 & 0 & 0 & -1 & 0 & -1 & -1 & 0 & 3
\end{bmatrix}
$$

(a) (b)

Figure 3

When we evaluate the $(7,3)$- and $(2,9)$-cofactors of $D(G) - A(G)$, using the MAPLE code shown in Figure 4, we see that the Petersen graph has 2000 nonidentical spanning trees—and we have demonstrated this by computing two cofactors, although only one need to be computed. We certainly would not want to solve this problem by *simply* drawing the 2000 nonidentical spanning trees.

```
⟩ r1  := [3, −1, 0, 0, −1, −1, 0, 0, 0, 0] :

⟩ r2  := [−1, 3, −1, 0, 0, 0, −1, 0, 0, 0] :

⟩ r3  := [0, −1, 3, −1, 0, 0, 0, −1, 0, 0] :
                                                ⟩ q2  := array([r.(1..10)]) :
⟩ r4  := [0, 0, −1, 3, −1, 0, 0, 0, −1, 0] :
                                                ⟩ (−1)^(7 + 3) * det(minor(q2, 7, 3));
⟩ r5  := [−1, 0, 0, −1, 3, 0, 0, 0, 0, −1] :
                                                .....................................
⟩ r6  := [−1, 0, 0, 0, 0, 3, 0, −1, −1, 0] :   ⟨                              2000

⟩ r7  := [0, −1, 0, 0, 0, 0, 3, 0, −1, −1] :   ⟩ (−1)^(2 + 9) * det(minor(q2, 2, 9));
                                                .....................................
⟩ r8  := [0, 0, −1, 0, 0, −1, 0, 3, 0, −1] :   ⟨                              2000

⟩ r9  := [0, 0, 0, −1, 0, −1, −1, 0, 3, 0] :

⟩ r10 := [0, 0, 0, 0, −1, 0, −1, −1, 0, 3] :
```

Figure 4

Directed Graphs and Rooted Spanning Trees

In this section we provide a directed version of our earlier work. Here we reexamine a directed graph, or digraph, $G = (V, E)$, where the nonempty set V contains the vertices of G and the set E contains the directed edges in G. Such a directed edge has the form (x, y) where $x, y \in V$ with $x \neq y$ and $(x, y) \neq (y, x)$. Figure 5(a) provides an example of a directed graph $G = (V, E)$, where $V = \{1, 2, 3, 4, 5\}$ and $E = \{(1, 2), (1, 4), (1, 5), (2, 3), (2, 4), (3, 1), (4, 3), (5, 4)\}$. The subgraphs in Figures 5(b) and (c) provide two of the nonidentical directed spanning trees of G—rooted at vertex 1. How many such spanning trees are there?

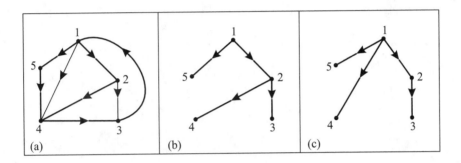

Figure 5

The adjacency matrix of G is now defined by $A(G) = (a_{ij})_{5 \times 5}$, where $a_{ij} = 1$ if $(i, j) \in E$, and $a_{ij} = 0$, otherwise. The degree matrix $D(G)$ from the previous section now becomes the indegree matrix; that is, $D(G) = (d_{ij})_{5 \times 5}$, where

$$d_{ij} = \begin{cases} \text{the number of edges that terminate at vertex } i, & \text{for } i = j, \\ 0, & \text{for } i \neq j. \end{cases}$$

Comparable to the Matrix-Tree Theorem, here we find that the number of nonidentical directed spanning trees for G—rooted at vertex i—is *the* value of the cofactor associated with the minor of $D(G) - A(G)$ obtained by deleting row i and column i from $D(G) - A(G)$. For the directed graph in Figure 5(a) one finds that

$$A(G) = \begin{array}{c} \\ (1) \\ (2) \\ (3) \\ (4) \\ (5) \end{array} \begin{array}{ccccc} (1) & (2) & (3) & (4) & (5) \\ \left[\begin{array}{ccccc} 0 & 1 & 0 & 1 & 1 \\ 0 & 0 & 1 & 1 & 0 \\ 1 & 0 & 0 & 0 & 0 \\ 0 & 0 & 1 & 0 & 0 \\ 0 & 0 & 0 & 1 & 0 \end{array}\right] \end{array} \quad \text{and} \quad D(G) = \begin{array}{c} \\ (1) \\ (2) \\ (3) \\ (4) \\ (5) \end{array} \begin{array}{ccccc} (1) & (2) & (3) & (4) & (5) \\ \left[\begin{array}{ccccc} 1 & 0 & 0 & 0 & 0 \\ 0 & 1 & 0 & 0 & 0 \\ 0 & 0 & 2 & 0 & 0 \\ 0 & 0 & 0 & 3 & 0 \\ 0 & 0 & 0 & 0 & 1 \end{array}\right] \end{array}.$$

To answer the question posed here we examine the cofactor for the minor obtained when the first row and first column are deleted from $D(G) - A(G)$. The computation in Figure 6 tells us that for the directed graphs in question, the number of nonidentical directed spanning trees—rooted at 1—is

> q^3 := matrix$(5, 5, [1, -1, 0, -1, -1, 0, 1, -1, -1, 0, -1, 0, 2, 0, 0, 0, 0, -1, 3, 0, 0, 0, 0,$
> $-1, 1])$:

> $(-1)\hat{\ }(1+1) * \det(\mathrm{minor}(q3, 1, 1));$

. . . .

<   6

Figure 6

$$(-1)^{1+1}\begin{vmatrix} 1 & -1 & -1 & 0 \\ 0 & 2 & 0 & 0 \\ 0 & -1 & 3 & 0 \\ 0 & 0 & -1 & 1 \end{vmatrix} = 6.$$

The reader who wishes to see more on the material presented in this section should examine Chapter 3 of [3].

Spanning Trees for the Wheel Graphs

This final section will provide another opportunity to apply what we have learned via the Matrix-Tree Theorem. It should appeal to those readers who have had a first course in discrete or combinatorial mathematics where some of the properties of the Fibonacci and Lucas numbers were studied.

In Figures 7(a), (b), and (c) we have drawn the graphs of the wheels W_3, W_4, and W_5, respectively. In general, for $n \geq 3$, we can describe the undirected wheel graph $W_n = (V, E)$ as the graph with $V = \{1, 2, 3, \ldots, n, n+1\}$ and $E = \{\{2, 3\}, \{3, 4\}, \ldots, \{n, n+1\}, \{n+1, 2\}\} \cup \{\{1, 2\}, \{1, 3\}, \ldots, \{1, n\}, \{1, n+1\}\}$.

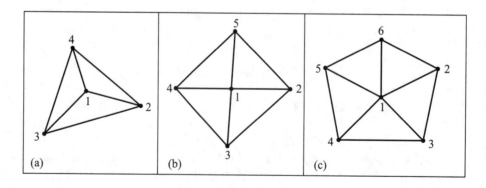

Figure 7

Using the ideas and notation given in the first section, we find $D(G) - A(G)$ for W_3, W_4, and W_5 to be respectively

$$\begin{bmatrix} 3 & -1 & -1 & -1 \\ -1 & 3 & -1 & -1 \\ -1 & -1 & 3 & -1 \\ -1 & -1 & -1 & 3 \end{bmatrix}, \quad \begin{bmatrix} 4 & -1 & -1 & -1 & -1 \\ -1 & 3 & -1 & 0 & -1 \\ -1 & -1 & 3 & -1 & 0 \\ -1 & 0 & -1 & 3 & -1 \\ -1 & -1 & 0 & -1 & 3 \end{bmatrix}, \quad \text{and}$$

$$\begin{bmatrix} 5 & -1 & -1 & -1 & -1 & -1 \\ -1 & 3 & -1 & 0 & 0 & -1 \\ -1 & -1 & 3 & -1 & 0 & 0 \\ -1 & 0 & -1 & 3 & -1 & 0 \\ -1 & 0 & 0 & -1 & 3 & -1 \\ -1 & -1 & 0 & 0 & -1 & 3 \end{bmatrix}.$$

Evaluating the $(1, 1)$–cofactor for each of these matrices, as shown by the inputs and outputs in Figure 8, we find that the number of nonidentical spanning trees for each of W_3, W_4 and

```
⟩ q4 := matrix(4, 4, [3, −1, −1, −1, −1, 3, −1, −1, −1, −1, 3, −1, −1, −1, −1, 3]) :

⟩ (−1)^(1 + 1) * det(minor(q4, 1, 1));
..........................................................................
⟨                                       16

⟩ q5 := matrix(5, 5, [4, −1, −1, −1, −1, −1, 3, −1, 0, −1, −1, −1, 3, −1, 0, −1, 0, −1, 3,
    −1, −1, −1, 0, −1, 3]) :

⟩ (−1)^(1 + 1) * det(minor(q5, 1, 1));
..........................................................................
⟨                                       45

⟩ q6 := matrix(6, 6, [5, −1, −1, −1, −1, −1, −1, 3, −1, 0, 0, −1, −1, −1, 3, −1, 0, 0, −1,
    0, −1, 3, −1, 0, −1, 0, 0, −1, 3, −1, −, 1, −1, 0, 0, −1, 3]) :

⟩ (−1)^(1 + 1) * det(minor(q6, 1, 1));
..........................................................................
⟨                                       121
```

Figure 8

W_5 is 16, 45, and 121, respectively. In order to conjecture a formula for the number $t(W_n)$ of nonidentical spanning trees for W_n in general, we observe that

$$t(W_3) = 16 = 18 - 2 = L_6 - 2$$
$$t(W_4) = 45 = 47 - 2 = L_8 - 2$$
$$t(W_5) = 121 = 123 - 2 = L_{10} - 2,$$

where L_n denotes the nth Lucas number.

Note: The Lucas numbers are defined recursively by $L_0 = 2, L_1 = 1$, and $L_n = L_{n-1} + L_{n-2}, n \geq 2$. The first 11 Lucas numbers are 2, 1, 3, 4, 7, 11, 18, 29, 47, 76, 123.

It can be shown that in general

$$t(W_n) = L_{2n} - 2, \quad n \geq 3.$$

Sometimes one finds the result expressed as

$$t(W_n) = F_{2n+2} - F_{2n-2} - 2, \quad n \geq 3,$$

where F_n denotes the nth Fibonacci number.

Note: The Fibonacci numbers are defined recursively by $F_0 = 0, F_1 = 1$, and $F_n = F_{n-1} + F_{n-2}, n \geq 2$. The first 11 Fibonacci numbers are 0, 1, 1, 2, 3, 5, 8, 13, 21, 34, 55.

The reader who wishes to see more in the way of a formal proof for the formula for $t(W_n)$ should examine [7].

References

1. Anton, Howard, *Elementary Linear Algebra*, 6th ed., New York: Wiley, 1991.

2. Behzad, Mehdi, Chartrand, Gary, and Leniak-Foster, Linda. *Graphs & Digraphs*, Belmont, CA: Wadsworth, 1979.

3. Gould, Ronald, *Graph Theory*, Reading, MA: Benjamin Cummings, 1988.

4. Grimaldi, Ralph P., *Discrete and Combinatorial Mathematics*, 3rd ed., Reading, MA: Addison Wesley, 1994.

5. Harary, Frank, *Graph Theory*, Reading, MA: Addison Wesley, 1969.

6. Lay, David C., *Linear Algebra and Its Applications*, Reading, MA: Addison Wesley, 1994.

7. Sedlacek, J., "On the Skeletons of a Graph or Digraph," *Proceedings of the Calgary International Conference on Combinatorial Structures and Their Applications* (1969), 387–391.

Ralph P. Grimaldi and Robert J. Lopez
Rose-Hulman Institute of Technology
Contributed

Algebraic Integers and Tensor Products of Matrices

A complex number is called an *algebraic integer* if it is a zero of a monic (i.e., the coefficient of the term with the highest exponent is one) polynomial with integer coefficients. Let $\mathbb{C}$ denote the complex numbers. A set $S \subseteq \mathbb{C}$ is called a *subring* if $1 \in S$, and for every $a, b \in S, a + b \in S$ and $a \cdot b \in S$. It is well-known that the set of algebraic integers forms a subring of the complex numbers (see [3]). The aim of the present note is to show that this result follows easily from basic facts about tensor products of matrices.

First we will need some notation. The set of $n \times n$ matrices over a subring R is denoted by $M_n(R)$. The set of polynomials with integer coefficients is denoted by $Z[x]$, x an indeterminate. If $A \in M_n(R)$, then $p_A(x) = \det(xI_n - A)$ is the *characteristic polynomial* of A. Since the operations involved in computing the determinant of A are multiplication and addition, it is readily verified that if $A \in M_n(Z)$, then $p_A(x)$ is a monic polynomial in $Z[x]$. Recall that if λ is a zero of $p_A(x)$, then λ is called an *eigenvalue* of A.

The following result will establish the relationship between algebraic integers and matrices.

Lemma 1. *Let λ be a complex number. Then λ is an algebraic integer if and only if λ is an eigenvalue of a square matrix with integer coefficients.*

Proof. If λ is an eigenvalue of $A \in M_n(Z)$, then $p_A(x)$ is a monic polynomial in $Z[x]$, and $p_A(\lambda) = 0$. Thus λ is an algebraic integer. To prove the converse, assume that $f(\lambda) = 0$, where $f(x)$ is a monic polynomial in $Z[x]$, say $f(x) = x^n + c_1 x^{n-1} + \cdots + c_{n-1} x + c_n$. Let A be the companion matrix of $f(x)$, i.e.

$$
A = \begin{bmatrix}
0 & 1 & 0 & \cdots & 0 \\
0 & 0 & 1 & \cdots & 0 \\
\vdots & \vdots & \ddots & \ddots & \vdots \\
0 & 0 & \cdots & 0 & 1 \\
-c_n & -c_{n-1} & \cdots & -c_2 & -c_1
\end{bmatrix}.
$$

Expanding $p_A(x)$ by the first column and using a simple induction argument, it follows that $p_A(x) = f(x)$, and this fact can be found in [1; pp 230–231]. Therefore λ is an eigenvalue of A and $A \in M_n(Z)$. ∎

Definition 2. If $A = [a_{ij}]$ is an $m \times n$ matrix and $B = [b_{ij}]$ is a $p \times q$ matrix then the *Kronecker* or *tensor product* of A and B, denoted $A \otimes B$, is the $mp \times nq$ matrix defined as follows:

$$A \otimes B = \begin{bmatrix} a_{11}B & \cdots & a_{1n}B \\ \vdots & \ddots & \vdots \\ a_{m1}B & \cdots & a_{mn}B \end{bmatrix}.$$

Note that if A and B have integer entries, then so does $A \otimes B$. From the above definition it is easy to verify that if $A, B, C, D \in M_n(R)$, then

$$(A \otimes B)(C \otimes D) = AC \otimes BD.$$

Vectors in C^n may be identified with $n \times 1$ matrices over C, so the above definition may be used to define tensor products $x \otimes y$ with $x \in C^n$ and $y \in C^m$. The following proposition establishes a part of the well-known results that identify the eigenvalues of tensor products see, e.g., [2; pp 242–245]. We provide an easy proof for completeness.

Proposition 3. *Let $A \in M_n(Z), B \in M_m(Z)$, let λ be an eigenvalue of A, and let μ be an eigenvalue of B. Then*

(a) $\lambda\mu$ is an eigenvalue of $A \otimes B$,

(b) $\lambda + \mu$ is an eigenvalue of $(A \otimes I_m) + (I_n \otimes B)$.

Proof. There exist non-zero vectors $x \in C^n$ and $y \in C^m$ such that $Ax = \lambda x$ and $By = \mu y$. It follows immediately that

$$(A \otimes B)(x \otimes y) = \lambda\mu(x \otimes y),$$

and

$$((A \otimes I_m) + (I_n \otimes B))(x \otimes y) = (\lambda + \mu)(x \otimes y).$$

This proves the proposition. ∎

Notice that if $A \in M_n(Z), B \in M_m(Z)$, then $A \otimes B \in M_{mn}(Z)$. Hence, if two algebraic integers, α, β are given, the tensor product is a useful tool for determining the existence of an integral matrix, for which $\alpha \cdot \beta$ or $\alpha + \beta$ is an eigenvalue. We summarize this in the following theorem.

Theorem 4. *The set of algebraic integers forms a subring of the ring of complex numbers.*

The proof follows directly from Lemma 2 and Proposition 4.

References

1. K. Hoffman and R. Kunze, *Linear Algebra*, Prentice Hall, New Jersey, 1971.

2. R.A. Horn and C.R. Johnson, *Topics in Matrix Analysis*, Cambridge University Press, New York, 1991.

3. H.L. Montgomery, I. Niven and H.S. Zuckerman, *An Introduction to the Theory of Numbers*, John Wiley and Sons, Inc., New York, 1991.

Shaun M. Fallat
University of Victoria
Edited
Crux Mathematicorum **22** (1996), 341–343

On Regular Markov Chains

The following basic theorem concerning regular Markov Chains is often presented in courses in finite mathematics.

Theorem. *Let P be the transition matrix for a regular Markov chain, then*

(i) $P^n \to W$ as $n \to \infty$,

(ii) each column of W is the same probability vector w which has positive components,

(iii) if p is a probability vector, $p^n p \to w$,

(iv) $w = Pw$.

Since, generally speaking, students in such courses know little mathematics beyond high school algebra, the proof of the theorem is usually omitted. In this note we present an elementary proof in the case when P is a 2×2 matrix.

Recall that a probability vector has nonnegative components that sum to 1, a transition matrix is a square matrix whose columns are probability vectors, and a transition matrix for a regular Markov chain has the property that some power of the matrix has all positive entries.

A regular 2×2 transition matrix can be written in the form

$$P = \begin{bmatrix} 1-a & b \\ a & 1-b \end{bmatrix},$$

where $0 < a \le 1, 0 < b \le 1, a+b < 2$. Before computing powers of P, we write $P = I + S$, where

$$S = \begin{bmatrix} -a & b \\ a & -b \end{bmatrix}.$$

Letting $r = -(a+b)$, we find $S^2 = rS$, and $S^n = r^{n-1}S$ for $n \ge 2$. Computing the powers of P we find

$$P^2 = (I + S)^2 = I + 2S + S^2 = I + [1 + (1+r)] S,$$

$$P^3 = I + [1 + (1+r) + (1+r)^2] S,$$

$$\vdots$$

$$P^n = I [1 + (1+r) + \cdots + (1+r)^{n-1}] S = I + \frac{1 - (1+r)^n}{1 - (1+r)} S.$$

235

Since $-1 < 1 + r < 1$, we find

$$P^n \to W = I + \frac{1}{a+b}S = \begin{bmatrix} \dfrac{b}{a+b} & \dfrac{b}{a+b} \\[2mm] \dfrac{a}{a+b} & \dfrac{a}{a+b} \end{bmatrix}.$$

This proves parts (i) and (ii) of the theorem; the other parts follow easily.

Finally we note that another way of computing P^n is to use the binomial theorem:

$$P^n = (I+S)^n = \sum_{i=0}^{n} \binom{n}{i} S^i = I + \sum_{i=1}^{n} \binom{n}{i} S$$

$$= I + \frac{1}{r} \left[\sum_{i=1}^{n} \binom{n}{i} r^i \right] S$$

$$= I + \frac{1}{r} \left[(1+r)^n - 1 \right] S.$$

Nicholas J. Rose
North Carolina State University
Contributed

Integration by Matrix Inversion

The integration of several functions using differential operators was considered by Osborn [1]. The integration of these and certain other functions by matrix inversion can furnish an application of several aspects of matrix theory of interest to the student of matrix algebra.

Let V be the vector space of differentiable functions. Let the n-tuple f be a basis spanning a subspace S of V which is closed under differentiation. Then differentation comprises a linear transformation T of S into itself. If the matrix A represents T relative to f, then when A is nonsingular the elements of fA^{-1} yield antiderivatives of the elements of f.

To integrate $e^{ax} \sin bx$ and $e^{ax} \cos bx$ consider $f = (e^{ax} \sin bx, e^{ax} \cos bx)$. Then

$$fT = (ae^{ax} \sin bx + be^{ax} \cos bx, -be^{ax} \sin bx + ae^{ax} \cos bx)$$

and

$$A = \begin{pmatrix} a & -b \\ b & a \end{pmatrix}.$$

Furthermore

$$A^{-1} = \frac{1}{a^2 + b^2} \begin{pmatrix} a & b \\ -b & a \end{pmatrix}$$

and then

$$fA^{-1} = \left(\frac{e^{ax}}{a^2 + b^2} \left(a \sin bx - b \cos bx\right), \frac{e^{ax}}{a^2 + b^2} \left(b \sin bx + a \cos bx\right) \right)$$

yields antiderivatives of the elements of f.

To derive the formula

$$(1) \qquad \int x^n e^x dx = e^x \left[x^n - nx^{n-1} + n(n-1)x^{n-2} - \cdots + (-1)^n n! \right]$$

for positive integers n consider $f = (e^x, xe^x, x^2 e^x, \ldots, x^n e^x)$. Then

$$fT = \left(e^x, e^x + xe^x, \ldots, nx^{n-1} e^x + x^n e^x \right)$$

and there follows the interesting matrix

$$A = \begin{bmatrix} 1 & 1 & 0 & \cdot & \cdot & \cdot & 0 & 0 & 0 \\ 0 & 1 & 2 & \cdot & \cdot & \cdot & 0 & 0 & 0 \\ 0 & 0 & 1 & \cdot & \cdot & \cdot & 0 & 0 & 0 \\ \cdot & \cdot & \cdot & \cdot & \cdot & \cdot & \cdot & \cdot & \cdot \\ \cdot & \cdot & \cdot & \cdot & \cdot & \cdot & \cdot & \cdot & \cdot \\ \cdot & \cdot & \cdot & \cdot & \cdot & \cdot & \cdot & \cdot & \cdot \\ 0 & 0 & 0 & \cdot & \cdot & \cdot & 1 & n-1 & 0 \\ 0 & 0 & 0 & \cdot & \cdot & \cdot & 0 & 1 & n \\ 0 & 0 & 0 & \cdot & \cdot & \cdot & 0 & 0 & 1 \end{bmatrix}.$$

Since only an antiderivative of the last element of f is required, one is only interested in the last column of A^{-1}. Due to the peculiar form of A the inverse is easily deduced. One surmises that the last column of A^{-1} is the transpose of the row

$$(2) \qquad \left((-1)^n n!, (-1)^{n-1} n!, (-1)^{n-2} \frac{n!}{2!}, \dots , n(n-1), -n, 1 \right).$$

That this supposition is correct may be verified by induction on n. In this connection it is useful to consider the $(n+2)$ rowed matrix corresponding to A as a partitioned matrix containing A as a principal submatrix. Finally, one notes that multiplication of (2) by f yields the required formula (1).

Reference

1. Roger Osborn, Note on integration by operators, *American Mathematical Monthly*, vol. 64, 1957, p. 431.

William Swartz
Montana State College
American Mathematical Monthly **65** (1958), 282–283.

Some Explicit Formulas
for the Exponential Matrix

In a recent paper, E.J. Putzer [1] described two methods for calculating exponential matrices of the form e^{tA}, where t is a scalar and A is any square matrix. Putzer's methods are particularly useful in practice because they are valid for all square matrices A and require no preliminary transformations of any kind. All that is needed is the factorization of the characteristic polynomial of A, that is, a knowledge of the eigenvalues of A and their multiplicities. Both methods are based on the fact that e^{tA} is a polynomial in A whose coefficients are scalar functions of t that can be determined recursively by solving a simple system of first-order linear differential equations.

A purely algebraic method for computing e^{tA} is given by the Lagrange-Sylvester interpolation formula described on pp. 101–102 of Gantmacher's *Theory of Matrices* [2]. This formula requires knowledge of the factorization of the minimal polynomial of A and is usually more complicated than Putzer's methods.

Another algebraic method for computing e^{tA} was developed recently by R.B. Kirchner [3] who gave an explicit formula for calculating e^{tA} in terms of A and the factorization of the characteristic polynomial of A. Kirchner's method requires the inversion of a certain matrix polynomial $q(A)$, although as Kirchner points out, this inversion can sometimes be avoided.

General methods often have the disadvantage that they are not the simplest methods to use in certain special cases. The purpose of this note is to point out that explicit formulas for the polynomial e^{tA} can be obtained very easily (a) if all the eigenvalues of A are equal, (b) if all the eigenvalues of A are distinct, or (c) if A has two distinct eigenvalues, exactly one of which has multiplicity 1. We state these formulas in the following three theorems.

Theorem 1. If A is an $n \times n$ matrix with all its eigenvalues equal to λ, then we have

$$(1) \qquad e^{tA} = e^{\lambda t} \sum_{k=0}^{n-1} \frac{t^k}{k!} (A - \lambda I)^k.$$

Proof. Since the matrices $\lambda t I$ and $t(A - \lambda I)$ commute, we have

$$e^{tA} = e^{\lambda t I} e^{t(A - \lambda I)} = (e^{\lambda t} I) \sum_{k=0}^{\infty} \frac{t^k}{k!} (A - \lambda I)^k.$$

The Cayley-Hamilton Theorem implies that $(A - \lambda I)^k = 0$ for $k \geq n$, so the theorem is proved.

NOTE. If $(A - \lambda I)^m = 0$ for some $m < n$, then the same proof shows that we can replace $n - 1$ by $m - 1$ in the upper limit of summation in (1).

Formula (1) is precisely the result obtained by applying Putzer's second method or Kirchner's explicit formula. The foregoing proof seems to be the simplest and most natural way to derive this result.

Theorem 2. *If A is an $n \times n$ matrix with n distinct eigenvalues $\lambda_1, \lambda_2, \cdots, \lambda_n$, then we have*

$$e^{tA} = \sum_{k=1}^{n} e^{t\lambda_k} L_k(A),$$

where the $L_k(A)$ are Lagrange interpolation coefficients given by

$$L_k(A) = \prod_{\substack{j=1 \\ j \neq k}} \frac{A - \lambda_j I}{\lambda_k - \lambda_j} \quad for\ k = 1, 2, \cdots, n.$$

Proof. Although this theorem is a special case of the Lagrange-Sylvester interpolation formula, the following alternate proof may be of interest.

Define a matrix-valued function of the scalar t by the equation

$$(2) \qquad F(t) = \sum_{k=1}^{n} e^{t\lambda k} L_k(A).$$

To prove that $F(t) = e^{tA}$ we show that F satisfies the differential equation $F'(t) = AF(t)$ and the initial conditions $F(0) = I$. From (2) we see that

$$AF(t) - F'(t) = \sum_{k=1}^{n} e^{t\lambda_k}(A - \lambda_k I) L_k(A).$$

By the Cayley-Hamilton Theorem we have $(A - \lambda_k I)L_k(A) = 0$ for each k, so F satisfies the differential equation $F'(t) = AF(t)$. Since

$$F(0) = \sum_{k=1}^{n} L_k(A) = I,$$

this completes the proof.

NOTE. My colleague Professor John Todd points out that another simple proof of Theorem 2 can be based on the fact that F satisfies the functional equation $F(s+t) = F(s)F(t)$.

The next theorem treats the case when A has two distinct eigenvalues, exactly one of which has multiplicity of 1.

Theorem 3. *Let A be an $n \times n$ matrix $(n \geq 3)$ with two distinct eigenvalues λ and μ, where λ has multiplicity $n-1$ and μ has multiplicity 1. Then we have*

$$e^{tA} = e^{\lambda t} \sum_{k=0}^{n-2} \frac{t^k}{k!}(A - \lambda I)^k$$

$$+ \left\{ \frac{e^{\mu t}}{(\mu - \lambda)^{n-1}} - \frac{e^{\lambda t}}{(\mu - \lambda)^{n-1}} \sum_{k=0}^{n-2} \frac{t^k}{k!}(\mu - \lambda)^k \right\} (A - \lambda I)^{n-1}.$$

Proof. As in the proof of Theorem 1 we write

$$e^{tA} = e^{\lambda t} \sum_{k=0}^{\infty} \frac{t^k}{k!}(A - \lambda I)^k = e^{\lambda t} \sum_{k=0}^{n-2} \frac{t^k}{k!}(A - \lambda I)^k + e^{\lambda t} \sum_{k=n-1}^{\infty} \frac{t^k}{k!}(A - \lambda I)^k$$

$$= e^{\lambda t} \sum_{k=0}^{n-2} \frac{t^k}{k!}(A - \lambda I)^k + e^{\lambda t} \sum_{r=0}^{\infty} \frac{t^{n-1+r}}{(n-1+r)!}(A - \lambda I)^{n-1+r}.$$

Now we evaluate the series over r in closed form by using the Cayley-Hamilton Theorem. Since $A - \mu I = A - \lambda I - (\mu - \lambda)I$, we find $(A - \lambda I)^{n-1}(A - \mu I) = (A - \lambda I)^n - (\mu - \lambda)(A - \lambda I)^{n-1}$. The left member is 0 by the Cayley-Hamilton Theorem so

$$(A - \lambda I)^n = (\mu - \lambda)(A - \lambda I)^{n-1}.$$

Using this relation repeatedly we find

$$(A - \lambda I)^{n-1+r} = (\mu - \lambda)^r (A - \lambda I)^{n-1}.$$

Therefore the series over r becomes

$$\sum_{r=0}^{\infty} \frac{t^{n-1+r}}{(n-1+r)!}(\mu - \lambda)^r (A - \lambda I)^{n-1} = \frac{1}{(\mu - \lambda)^{n-1}} \sum_{k=n-1}^{\infty} \frac{t^k}{k!}(\mu - \lambda)^k (A - \lambda I)^{n-1}$$

$$= \frac{1}{(\mu - \lambda)^{n-1}} \left\{ e^{t(\mu - \lambda)} - \sum_{k=0}^{n-2} \frac{t^k}{k!}(\mu - \lambda)^k \right\} (A - \lambda I)^{n-1}.$$

This completes the proof of Theorem 3.

The explicit formula in Theorem 3 can also be deduced by applying Putzer's method or by using Kirchner's formula, but the details are much more complicated.

The explicit formulas in Theorems 1, 2, and 3 cover all matrices of order $n \leq 3$. Since the 3×3 case is often discussed in the classroom, the formulas in this case are listed below for easy reference.

1. If a 3×3 matrix A has eigenvalues $\lambda, \lambda, \lambda$, then

$$e^{tA} = e^{\lambda t} \left\{ I + t(A - \lambda I) + \frac{1}{2}t^2(A - \lambda I)^2 \right\}.$$

2. If a 3×3 matrix A has distinct eigenvalues λ, μ, ν, then

$$e^{tA} = e^{\lambda t}\frac{(A-\mu I)(A-\nu I)}{(\lambda-\mu)(\lambda-\nu)} + e^{\mu t}\frac{(A-\lambda I)(A-\nu I)}{(\mu-\lambda)(\mu-\nu)} + e^{\nu t}\frac{(A-\lambda I)(A-\mu I)}{(\nu-\lambda)(\nu-\mu)}.$$

3. If a 3×3 matrix A has eigenvalues λ, λ, μ, with $\lambda \neq \mu$, then

$$e^{tA} = e^{\lambda t}\{I + t(A-\lambda I)\} + \frac{e^{\mu t} - e^{\lambda t}}{(\mu-\lambda)^2}(A-\lambda I)^2 - \frac{te^{\lambda t}}{\mu-\lambda}(A-\lambda I)^2.$$

References

1. E.J. Putzer, Avoiding the Jordan canonical form in the discussion of linear systems with constant coefficients, *American Mathematical Monthly* **73** (1966), 2–7.

2. F.R. Gantmacher, *The Theory of Matrices*, Volume I, Chelsea, New York, 1959.

3. R.B. Kirchner, An explicit formula for e^{At}, *American Mathematical Monthly* **74** (1967), 1200–1204.

T.M. Apostol
California Institute of Technology
American Mathematical Monthly **76** (1969), 289–292

Avoiding the Jordan Canonical Form in the Discussion of Linear Systems with Constant Coefficients

Consider the differential equation

$$(1) \qquad \dot{x} = Ax; \qquad x(0) = x_0; \qquad 0 \le t \le \infty,$$

where x and x_0 are n-vectors and A is an $n \times n$ matrix of constants. In this paper we present two methods, believed to be new, for explicitly writing down the solution of (1) without making any preliminary transformations. This is particularly useful, both for teaching and applied work, when the matrix A cannot be diagonalized, since the necessity of discussing or finding the Jordan Canonical Form (J.C.F.) of A is completely bypassed.

If e^{At} is defined as usual by a power series it is well known (see [1]) that the solution of (1) is

$$x = e^{At} x_0,$$

so the problem is to calculate the function e^{At}. In [2], this is done via the J.C.F. of A. In [1], it is shown how the J.C.F. can be bypassed by a transformation which reduces A to a triangular form in which the off diagonal elements are arbitrarily small. While this approach permits a theoretical discussion of the form $\exp\{At\}$ and its behavior as t becomes infinite (Theorem 7 of [1]), it is not intended as a practical method for calculating the function. The following two theorems suggest an alternate approach which can be used both for calculation and for expository discussion. It may be noted that the formula of Theorem 2 is simpler than that of Theorem 1 since the r_i are easier to calculate than the q_i.

In order to state Theorem 1 simply, it will be convenient to introduce some notation.

Let A be an $n \times n$ matrix of constants, and let

$$f(\lambda) \equiv |\lambda I - A| \equiv \lambda^n + c_{n-1}\lambda^{n-1} + \cdots + c_1 \lambda + c_0$$

be the characteristic polynomial of A. Construct the scalar function $z(t)$ which is the solution of the differential equation

$$(2) \qquad z^{(n)} + c_{n-1} z^{(n-1)} + \cdots + c_1 \dot{z} + c_0 z = 0$$

with initial conditions

$$(3) \qquad z(0) = \dot{z}(0) = \cdots = z^{(n-2)}(0) = 0; \qquad z^{(n-1)}(0) = 1.$$

We observe at this point that regardless of the multiplicities of the roots of $f(\lambda) = 0$, once these are obtained it is trivial to write down the general solution of (2). Then one solves a single set of linear algebraic equations to satisfy the initial conditions (3). Since the right member of each of these equations is zero except for the last, solving them entails only finding the cofactors of the elements of the *last row* of the associated matrix. It is not necessary to invert the matrix itself. For teaching purposes, the point is that the form of the general solution of (2) can be obtained quickly and easily by elementary methods.

Now define

$$Z(t) = \begin{bmatrix} z(t) \\ \dot{z}(t) \\ \vdots \\ z^{(n-1)}(t) \end{bmatrix}, \quad \text{and} \quad C = \begin{bmatrix} c_1 & c_2 \cdots c_{n-1} & 1 \\ c_2 & c_3 \cdots 1 & \\ \vdots & \ddots & 0 \\ c_{n-1} & 1 & \\ 1 & & \end{bmatrix}$$

We then have the following

Theorem 1.

$$(4) \qquad\qquad e^{At} = \sum_{j=0}^{n-1} q_j(t) A^j,$$

where $q_0(t), \cdots, q_{n-1}(t)$ *are the elements of the column vector*

$$(5) \qquad\qquad q(t) = CZ(t).$$

Before we prove this, a remark is in order as to what happens when $f(\lambda)$ has multiple roots but the minimal polynomial of A has distinct factors so that A can in fact be diagonalized. It appears at first glance that our formula will contain powers of t, yet we know this cannot be the case. What occurs, of course, is that the powers of t in (4) just cancel each other out. This is the nice feature of formula (4); it is true for *all* matrices A, and we never have to concern ourselves about the nature of the minimal polynomial of A, or its J.C.F., and no preliminary transformations of any kind need be made.

Proof. We will show that if

$$(6) \qquad\qquad \Phi(t) = \sum_{j=0}^{n-1} q_j(t) A^j$$

then $d\Phi/dt = A\Phi$ and $\Phi(0) = I$, so $\Phi(t) = e^{At}$. Since only $q_0(t)$ involves $z^{(n-1)}(t)$, $q_j(0) = 0$ for $j \geq 1$. Clearly, $q_0(0) = 1$. Thus, $\Phi(0) = I$.

Now consider $(d\Phi/dt) - A\Phi$. Differentiating (6), and applying the Hamilton-Cayley Theorem

$$A^n + \sum_{j=0}^{n-1} c_j A^j = 0,$$

we obtain

$$\frac{d\Phi}{dt} - A\Phi = (\dot{q}_0 + c_0 q_{n-1}) + \sum_{j=0}^{n-1} (\dot{q}_j - q_{j-1} + c_j q_{n-1}) A^j.$$

It will suffice, therefore, to show that

$$\dot{q}_0(t) \equiv -c_0 q_{n-1}(t),$$
$$\dot{q}_j(t) \equiv q_{j-1}(t) - c_j q_{n-1}(t) \qquad j = 1, \cdots, (n-1).$$

From the definition (5),

$$(7) \qquad\qquad q_j(t) \equiv \sum_{k=1}^{n-j-1} c_{k+j} z^{(k-1)} + z^{(n-j-1)}.$$

Therefore $\dot{q}_j(t) \equiv \sum_{k=1}^{n-j-1} c_{k+j} z^{(k)} + z^{(n-j)}$.

But $q_{n-1} \equiv z$, so we have

$$(8) \qquad\qquad \dot{q}_j + c_j q_{n-j} \equiv \sum_{k=0}^{n-j-1} c_{k+j} z^{(k)} + z^{(n-j)} \quad \text{for } j = 0, 1, \cdots, n-1.$$

If $j = 0$, this yields

$$\dot{q}_0 + c_0 q_{n-1} \equiv \sum_{k=0}^{n-1} c_k z^{(k)} + z^{(n)}$$

which is zero because of (2).

If $j \geq 1$, replace j by $j-1$ in (7) and change the summation index from k to $k+1$ to get

$$(9) \qquad\qquad q_{j-1}(t) \equiv \sum_{k=0}^{n-j-1} c_{k+j} z^{(k)} + z^{(n-j)}.$$

Comparing (9) and (8) we have $\dot{q}_j + c_j q_{n-1}(t) \equiv q_{j-1}(t)$ for $j = 1, 2, \cdots, n-1$.

Students will want to see the formula (4) *derived*. One merely presents the proof backward, beginning with the observation that because of the Hamilton-Cayley theorem, $\exp\{At\}$ should be expressible in the form (6). Regarding the q_j as unknowns, and applying the differential equation which $\exp\{At\}$ satisfies, leads directly to (4).

A second explicit formula for $\exp\{At\}$, which also holds for all matrices A, is the following: Let $\lambda_1, \lambda_2, \cdots, \lambda_n$ be the eigenvalues of A in some arbitrary but specified order. These are not necessarily distinct. Then

Theorem 2. $e^{At} = \sum_{j=0}^{n-1} r_{j+1}(t) P_j$, *where*

$$P_0 = I; \quad P_j = \prod_{k=1}^{j} (A - \lambda_k I), \qquad (j = 1, \cdots, n),$$

and $r_1(t), \cdots, r_n(t)$ *is the solution of the triangular system*

$$\begin{cases} \dot{r}_1 = \lambda_1 r_1 \\ \dot{r}_j = r_{j-1} + \lambda_j r_j & (j = 2, \cdots n) \\ r_1(0) = 1; \ r_j(0) = 0 & (j = 2, \cdots n). \end{cases}$$

Proof. Let

(10)
$$\Phi(t) \equiv \sum_{j=0}^{n-1} r_{j+1}(t) P_j$$

and define $r_0(t) \equiv 0$. Then from (10) and the equations satisfied by the $r_j(t)$ we have, after collecting terms in r_j,

$$\dot{\Phi} - \lambda_n \Phi = \sum_{0}^{n-2} \left[P_{j+1} + (\lambda_{j+1} - \lambda_n) P_j \right] r_{j+1}.$$

Using $P_{j+1} \equiv (A - \lambda_{j+1} I) P_j$ in this gives

$$\begin{aligned} \dot{\Phi} - \lambda_n \Phi \ &= (A - \lambda_n I)(\Phi - r_n(t) P_{n-1}) \\ &= (A - \lambda_n I)\, \Phi - r_n(t) P_n. \end{aligned}$$

But $P_n \equiv 0$ from the Hamilton-Cayley Theorem, so $\dot{\Phi} = A\Phi$. Then since $\Phi(0) = I$ it follows that $\Phi(t) = e^{At}$.

Example. If one desires a numerical example for class presentation an appropriate matrix A can be prepared in advance by arbitrarily choosing a set of eigenvalues, a Jordan Canonical Form J and a nonsingular matrix S and calculating

$$A = SJS^{-1}.$$

Then beginning with A, one simply calculates the set $\{q_i(t)\}$ and/or $\{r_i(t)\}$. Consider the case of a 3×3 matrix having eigenvalues (λ, λ, μ). There are two subcases; the one in which the normal form of A is diagonal, and the one in which it is not. These two subcases are taken care of automatically by the given formula for $\exp\{At\}$, and do not enter at all into the calculation of the $\{q_i\}$ or $\{r_i\}$.

As an example we will explicitly find the sets $\{q_i\}$ and $\{r_i\}$ for the case of a 3×3 matrix with eigenvalues $(\lambda, \lambda, \lambda)$. We note that aside from the trivial case in which the normal form (and hence A itself) is diagonal, there are two distinct nondiagonal normal forms that A may have. As above, these do not have to be treated separately.

From Theorem 1,

$$f(x) \equiv (x - \lambda)^3 = x^3 - 3\lambda x^2 + 3\lambda^2 x - \lambda^3$$

so $c_1 = 3\lambda^2, c_2 = -3\lambda$. Obviously $z(t) = (a_1 + a_2 t + a_3 t^2)e^{\lambda t}$. Applying the initial conditions to find the a_i yields $z = \frac{1}{2}t^2 e^{\lambda t}$, so

$$Z(t) = \frac{1}{2}e^{\lambda t}\begin{bmatrix} t^2 \\ \lambda t^2 + 2t \\ \lambda^2 t^2 + 4\lambda t + 2 \end{bmatrix}.$$

Then since

$$C = \begin{bmatrix} 3\lambda^2 & -3\lambda & 1 \\ -3\lambda & 1 & 0 \\ 1 & 0 & 0 \end{bmatrix}$$

$$q = CZ(t) = \frac{1}{2}e^{\lambda t}\begin{bmatrix} \lambda^2 t^2 - 2\lambda t + 2 \\ -2\lambda t^2 + 2t \\ t^2 \end{bmatrix}.$$

Thus

(11) $$e^{At} = \frac{1}{2}e^{\lambda t}\left\{(\lambda^2 t^2 - 2\lambda t + 2)I + (-2\lambda t^2 + 2t)A + t^2 A^2\right\}$$

for every 3×3 matrix A having all three eigenvalues equal to λ. The corresponding formula from Theorem 2 is obtained by solving the system

$$\begin{cases} \dot{r}_1 = \lambda r_1 \\ \dot{r}_2 = r_1 + \lambda r_2 \\ \dot{r}_3 = r_2 + \lambda r_3 \end{cases}$$

with the specified initial values. This immediately gives

$$r_1 = e^{\lambda t}; \qquad r_2 = te^{\lambda t}; \qquad r_3 = \frac{t^2}{2}e^{\lambda t}$$

so

(12) $$e^{At} = \frac{1}{2}e^{\lambda t}\left\{2I + 2t(A - \lambda I) + t^2(A - \lambda I)^2\right\}.$$

Of course, if we collect powers of A in (12) we will obtain (11).

References

1. R.E. Bellman, *Stability Theory of Differential Equations*, McGraw-Hill, New York, 1953.
2. E.A. Coddington and N. Levinson, *Theory of Ordinary Differential Equations*, McGraw-Hill, New York, 1955.

E.J. Putzer
North American Aviation Science Center
American Mathematical Monthly **73** (1966), 2–7

The Discrete Analogue of the Putzer Algorithm

In differential equations, the Putzer algorithm can be used to represent solutions of the form e^{At}. Here, we introduce an analogous algorithm to compute A^n and then use it to solve a difference equation.

For a $k \times k$ matrix A, denote the eigenvalues of A by $\lambda_1, \ldots, \lambda_k$. For any nonnegative integer n, we look for a representation of A^n in the form

$$A^n = \sum_{j=1}^{k} u_j(n) M_{j-1}, \tag{1}$$

where the $u_j(n)$ are scalar functions to be determined later, and

$$M_j = (A - \lambda_j I) M_{j-1}, \qquad M_0 = 1. \tag{2}$$

Thus, $M_j = \prod_{i=1}^{j} (A - \lambda_i I)$. Applying the Cayley-Hamilton Theorem, one may conclude that

$$M_k = \prod_{i=1}^{k} (A - \lambda_i I) = 0.$$

For $n = 0$ in formula (1) we have

$$A^0 = I = u_1(0) I + u_2(0) M_1 + \cdots + u_k(0) M_{k-1}.$$

This equation is satisfied if

$$u_1(0) = 1, \quad \text{and} \quad u_2(0) = u_3(0) = \cdots = u_k(0) = 0. \tag{3}$$

Now, from formula (1),

$$\sum_{j=1}^{k} u_j(n+1) M_{j-1} = A^{n+1} = AA^n$$

$$= A \left(\sum_{j=1}^{k} u_j(n) M_{j-1} \right) = \sum_{j=1}^{k} u_j(n) A M_{j-1}.$$

249

Using equation (2), one obtains

$$\sum_{j=1}^{k} u_j(n+1)M_{j-1} = \sum_{j=1}^{k} u_j(n)\left[M_j + \lambda_j M_{j-1}\right]. \tag{4}$$

Comparing the coefficients of M_j in equation (4), and applying condition (3), we have

$$\left. \begin{array}{l} u_1(n+1) = \lambda_1 u_1(n), \ u_1(0) = 1 \\ u_j(n+1) = \lambda_j u_j(n) + u_{j-1}(n), \ u_j(0) = 0, \quad j = 2, 3, \ldots, k \end{array} \right\}. \tag{5}$$

The solutions of Equations (5) are given by

$$u_1(n) = \lambda_1^n, \ u_j(n) = \sum_{i=0}^{n-1} \lambda_j^{n-1-i} u_{j-1}(i), \quad j = 2, 3, \ldots, k.$$

Equations (2) and (5) together constitute an algorithm for computing A^n.

Example. Find the solution of the difference system $x(n+1) = Ax(n)$, where

$$A = \begin{pmatrix} 4 & 1 & 2 \\ 0 & 2 & -4 \\ 0 & 1 & 6 \end{pmatrix}.$$

Solution: The eigenvalues of A may be obtained by solving the characteristic equation $\det(A - \lambda I) = 0$. Now

$$\det \begin{pmatrix} 4-\lambda & 1 & 2 \\ 0 & 2-\lambda & -4 \\ 0 & 1 & 6-\lambda \end{pmatrix} = (4-\lambda)(\lambda-4)^2 = 0.$$

Hence, the eigenvalues of A are $\lambda_1 = \lambda_2 = \lambda_3 = 4$. So

$$M_0 = I, \ M_1 = 4I = \begin{pmatrix} 0 & 1 & 2 \\ 0 & -2 & -4 \\ 0 & 1 & 2 \end{pmatrix}, \ \text{and } M_2 = (A-4I)M_1 = \begin{pmatrix} 0 & 0 & 0 \\ 0 & 0 & 0 \\ 0 & 0 & 0 \end{pmatrix}.$$

Now,

$$u_1(n) = 4^n$$

$$u_2(n) = \sum_{i=0}^{n-1} \left(4^{n-1-i}\right)\left(4^i\right) = n\left(4^{n-1}\right)$$

$$u_3(n) = \sum_{i=0}^{n-1} 4^{n-1-i}\left(i4^{i-1}\right) = 4^{n-2}\sum_{i=0}^{n-1} i = \frac{n(n-1)}{2}4^{n-2}.$$

Using equation (1), we have

$$A^n = 4n \begin{pmatrix} 1 & 0 & 0 \\ 0 & 1 & 0 \\ 0 & 0 & 1 \end{pmatrix} + n4^{n-1} \begin{pmatrix} 0 & 1 & 2 \\ 0 & -2 & -4 \\ 0 & 1 & 2 \end{pmatrix} + \frac{n(n-1)}{2} 4^{n-2} \begin{pmatrix} 0 & 0 & 0 \\ 0 & 0 & 0 \\ 0 & 0 & 0 \end{pmatrix}$$

$$= \begin{pmatrix} 4^n & n\,4^{n-1} & 2n\,4^{n-1} \\ 0 & 4^n - 2n\,4^{n-1} & -n\,4^n \\ 0 & n\,4^{n-1} & 4^n + 2n\,4^{n-1} \end{pmatrix}.$$

The solution of the difference equation is given by

$$x(n) = A^n x(0) = \begin{pmatrix} 4^n x_1(0) + n\,4^{n-1} x_2(0) + 2n\,4^{n-1} x_3(0) \\ (4^n - 2n\,4^{n-1}) x_2(0) - n\,4^n x_3(0) \\ n\,4^{n-1} x_2(0) + (4^n + 2n\,4^{n-1}) x_3(0) \end{pmatrix}$$

where $x(0) = (x_1(0), x_2(0), x_3(0))^T$.

Exercises

In Exercises 1–4, use the discrete Putzer algorithm to evaluate A^n.

1. $A = \begin{pmatrix} 1 & 1 \\ -2 & 4 \end{pmatrix}$
   ans. $\begin{pmatrix} 2^{n+1} - 3^n & 3^n - 2^n \\ 2^{n+1} - 2(3^n) & 2(3^n) - 2^n \end{pmatrix}$

2. $A = \begin{pmatrix} -1 & 2 \\ 3 & 0 \end{pmatrix}$

3. $A = \begin{pmatrix} 1 & 2 & -1 \\ 1 & 0 & 1 \\ 4 & -4 & 5 \end{pmatrix}$
   ans. $\begin{pmatrix} 2^{n+1} - 3 & -2 + 2^{n+1} & \frac{1}{2} - \frac{1}{2}3^n \\ -2 + 3^n & 2 - 2^n & -\frac{1}{2} - \frac{1}{2}3^n \\ -2^{n+2} + 4(3^n) & 4 - 2^{n+2} & -1 + 2(3^n) \end{pmatrix}$

4. $A = \begin{pmatrix} 2 & 1 & 0 \\ 0 & 2 & 1 \\ 0 & 0 & 2 \end{pmatrix}$

5. Solve the system
   $$x_1(n+1) = -x_1(n) + x_2(n), \quad x_1(0) = 1,$$
   $$x_2(n+1) = 2x_2(n), \quad x_2(0) = 2.$$
   ans. $\begin{pmatrix} \frac{1}{3}\left[2^{n+1} + (-1)^n\right] \\ 2^{n+1} \end{pmatrix}$

6. Solve the system

$$x_1(n+1) = x_2(n),$$
$$x_2(n+1) = x_3(n,),$$
$$x_3(n+1) = 2x_1(n) - x_2(n) + x_3(n).$$

7. Solve the system

$$x(n+1) = \begin{pmatrix} 1 & -2 & -2 \\ 0 & 0 & -1 \\ 0 & 2 & 3 \end{pmatrix} x(n), \ x(0) = \begin{pmatrix} 1 \\ 1 \\ 0 \end{pmatrix}. \qquad \text{ans.} \begin{pmatrix} 3 - 2^{n+1} \\ 2\left(1 - 2^{n-1}\right) \\ 2\left(-1 + 2^n\right) \end{pmatrix}$$

8. Solve the system

$$x(n+1) = \begin{pmatrix} 1 & 3 & 0 & 0 \\ 0 & 2 & 1 & -1 \\ 0 & 0 & 2 & 0 \\ 0 & 0 & 0 & 3 \end{pmatrix} x(n).$$

9. Verify that this matrix satisfies its characteristic equation (Cayley-Hamilton Theorem).

$$A = \begin{pmatrix} 2 & -1 \\ 1 & 3 \end{pmatrix}.$$

10. Let $\rho(A) = \max\{|\lambda| : \lambda \text{ is an eigenvalue of } A\}$. Suppose that $\rho(A) = \rho_0 < \beta$.

   (a) Show that $|u_j(n)| \le \dfrac{\beta^n}{(\beta - \rho_0)}, j = 1, 2, \dots, k$. [*Hint*: Use equation (5).]

   (b) Show that if $\rho_0 < 1$, then $u_j(n) \to 0$ as $n \to \infty$. Conclude that $A^n \to 0$ as $n \to \infty$.

   (c) If $\alpha < \min\{|\lambda| : \lambda \text{ is an eigenvalue of } A\}$, establish a lower bound for $|u_j(n)|$.

11. If a $k \times k$ matrix A has distinct eigenvalues $\lambda_1, \lambda_2, \dots, \lambda_k$, then one may compute $A^n, n \ge k$, using the following method. Let $p(\lambda)$ be the characteristic polynomial of A. Divide λ^n by $p(\lambda)$ to obtain $\lambda^n = p(\lambda)q(\lambda) + r(\lambda)$, where the remainder $r(\lambda)$ is a polynomial of degree at most $(k-1)$. Thus one may write $A^n = p(A)q(A) + r(A)$.

   (a) Show that $A^n = r(A) = a_0 I + a_1 A + a_2 A^2 + \cdots + a_{k-1}A^{k-1}$.

   (b) Show that $\lambda_1^n = r(\lambda_1), \lambda_2^n, \dots, \lambda_k^n = r(\lambda_k)$.

   (c) Use part (b) to find $a_0, a_1, \dots, a_{k-1}$.

12. Extend the method of Exercise 11 to the case of repeated roots. [*Hint*: If $\lambda_1 = \lambda_2 = \lambda$ and $\lambda^n = a_0 + a_1\lambda + a_2\lambda^2 + \cdots + a_{k-1}r^{k-1}$, differentiate to get another equation $n\lambda^{n-1} = a_1 + 2a_2\lambda + \cdots + (k-1)a_{k-1}\lambda^{k-2}$.]

13. Apply the method of Exercise 11 to find A^n for

   (a) $A = \begin{pmatrix} 1 & 1 \\ -2 & 4 \end{pmatrix}$,

   (b) $A = \begin{pmatrix} 1 & 2 & -1 \\ 1 & 0 & 1 \\ 4 & -4 & 5 \end{pmatrix}.$

ans. (a) $\begin{pmatrix} 2^{n+1} - 3^n & 3^n - 2^n \\ 2^{n+1} - 2(3^n) & 2(3^n) - 2^n \end{pmatrix}$ (b) Same as Exercise 3.

14. Apply the method of Exercise 13 to find A^n for

$$A = \begin{pmatrix} 4 & 1 & 2 \\ 0 & 2 & -4 \\ 0 & 1 & 6 \end{pmatrix}.$$

Saber Elaydi
Trinity University
Contributed

The Minimum Length of a Permutation as a Product of Transpositions

The intent of this note is to present results on minimal expressions of permutations as products of transpositions in a way that takes advantage of the linear algebra inherent in viewing permutations as orthogonal transformations. The known result that the minimum number of transpositions required is determined by the number of disjoint cycles in the permutation [1]-[4] is derived here using Gram-Schmidt orthogonalization. We rely only on elementary linear algebra in proofs and do not employ arguments based on case by case multiplication cycles. The transpositions that do occur in a minimal representation of a permutation will be shown to be linearly independent, in a sense to be made precise.

We view elements σ of the symmetric group S_n as acting on the Euclidean space $\mathbf{R}^n$ by $\sigma \mathbf{e}_i = \mathbf{e}_{\sigma(i)}$, where $\mathbf{e}_1, \mathbf{e}_2, \ldots, \mathbf{e}_n$ denotes the natural basis of $\mathbf{R}^n$. This action extends to a linear transformation that preserves the Euclidean length of vectors. Thus σ defines a real orthogonal linear transformation with $(\sigma x, \sigma y) = (x, y)$, for all x, y in $\mathbf{R}^n$, where (x, y) denotes the usual inner product.

In this setting, transpositions act as reflections through hyperplanes. In particular, the transposition $\tau = (i, j)$ can be identified with the reflection through the $(n-1)$-dimensional subspace (hyperplane) orthogonal to the vector $\mathbf{e}_i - \mathbf{e}_j$. This is so since τ sends the normal vector $\mathbf{e}_i - \mathbf{e}_j$ to its negative $\mathbf{e}_j - \mathbf{e}_i$, and also τ fixes pointwise the collection of vectors $\{\mathbf{e}_k | k \neq i, j\} \cup \{\mathbf{e}_i + \mathbf{e}_j\}$ which form a basis for the hyperplane.

By the fixed point space of σ, and of any linear transformation, we mean the set of vectors x in $\mathbf{R}^n$ with $\sigma(x) = x$. Provided it is non-trivial, this is the eigenspace of σ corresponding to the eigenvalue $\lambda = 1$ and we denote it V_σ. It is not difficult to see that the fixed point space is determined by the cycle structure of the permutation. If σ is written as a product of r disjoint cycles, including trivial cycles containing only one point, then V_σ is r-dimensional with each cycle of σ contributing a basis element in a natural way to V_σ.

For example, the permutation $\sigma = (2, 5, 3)(1, 6)$ in S_7 has a four-dimensional fixed point space in R^7 that has the vectors $\mathbf{e}_2 + \mathbf{e}_5 + \mathbf{e}_3, \mathbf{e}_1 + \mathbf{e}_6, \mathbf{e}_4$, and $\mathbf{e}_7$ for a basis.

We read products of permutations from right to left. For example, $(1, 2, 3, 4, 5) = (1, 5)(1, 4)(1, 3)(1, 2)$ expresses a five-cycle as a product of four transpositions. In like fashion, an s-cycle can be written using $s-1$ transpositions. Thus, a permutation in S_n consisting of r cycles can be written as a product of $n - r$ transpositions. A well-known result says that no fewer number of transpositions can be employed. The proof we give takes advantage of the linear algebraic bent with which we view permutations. Note that in counting cycles of a permutation we always include one-element cycles.

Theorem. *A permutation in S_n cannot be written as the product of fewer than $n - r$ transpositions, where r is the number of disjoint cycles in the permutation.*

Proof: Suppose σ in S_n is written as $\sigma = \tau_1\tau_2\cdots\tau_k$, where the τ_i's are transpositions. Viewing transpositions as reflections through hyperplanes, let v_i, $1, 2, \ldots, k$, be a non-zero vector orthogonal to the hyperplane determined by τ_i. The Gram-Schmidt orthogonalization process guarantees the existence of at least $n - k$ linearly independent vectors that are orthogonal to the subspace spanned by the v_i's. These $n-k$ vectors thus lie in the intersection of the k hyperplanes determined by the transpositions and are thus pointwise fixed by each of the transpositions. Thus these vectors are fixed by σ, and so $\dim V_\sigma \geq n - k$. But, $\dim V_\sigma = r =$ number of cycles in σ. The result $k \geq n - r$ follows. $\square$

Whenever $\sigma = \tau_1\tau_2\cdots\tau_k$, a product of transpositions, and k is the minimum number allowed by Theorem 1, we refer to this as a *minimal representation* of σ. Further, given the transposition $\tau = (i, j)$ in $S_n, i < j$, we call the vector $\mathbf{e}_i - \mathbf{e}_j$ in $\mathbf{R}^n$ the vector associated to τ.

For any σ in S_n, we always have the direct sum decomposition $R^n = V_\sigma \oplus V_\sigma^\perp$, where $V_\sigma^\perp$ denotes the orthogonal complement in $\mathbf{R}^n$ of the fixed point space V_σ. If $\sigma = \tau_1\tau_2\cdots\tau_k$ is a minimal representation, then $\dim V_\sigma = n - k$. Now the vectors $v_i, i = 1, 2, \ldots, k$, associated to the transpositions τ_i are normal vectors to hyperplanes H_i. Since τ_i fixes H_i pointwise, the intersection $\bigcap_{i=1}^k H_i$ is a subspace contained in V_σ. Elementary results concerning solution spaces of systems of equations show that $\dim\left(\bigcap_{i=1}^k H_i\right) \geq n - k$, with equality occurring exactly when the normal vectors $v_1, v_2, \ldots, v_k$ are linearly independent. We thus have the result that if $\sigma = \tau_1\tau_2\cdots\tau_k$ is a minimal representation, then the associated vectors $v_1, v_2, \ldots, v_k$ are linearly independent and form a basis for $v_\sigma^\perp$.

For example, $(1, 6)(3, 4)(4, 6)(1, 3)$ could not be a minimal representation due to the dependence relation $\mathbf{e}_1 - \mathbf{e}_6 = (\mathbf{e}_3 - \mathbf{e}_4) + (\mathbf{e}_4 - \mathbf{e}_6) - (\mathbf{e}_3 - \mathbf{e}_1)$.

Other reasonably intuitive results about minimal products of transpositions can be derived using this approach. For example, a minimal representation $\sigma = \tau_1\tau_2\cdots\tau_k$ must respect the cycle structure of σ. For, suppose that some transposition $\tau_i = (a, b)$ was such that a and b belonged to different cycles of σ. Then the vector $v = \mathbf{e}_a - \mathbf{e}_b$ would not have inner product zero with the vector $w = \sum \mathbf{e}_\alpha$, where α ranges through the elements of the cycle of σ containing a. But this contradicts our result that w is in V_σ, while the associated vector v is in $V_\sigma^\perp$. In particular, no transposition $\tau_i = (a, b)$ can have either a or b belonging to a trivial one-element cycle of σ.

References

1. Josef Denes, The representation of a permutation as the product of a minimal number of transpositions, and its connection with the theory of graphs, *Publ. Math. Institute Hung. Acad. Sci.* **4** (1959), 63–71.

2. O.P. Lossers, Solution to Problem E3058, *American Mathematical Monthly* **93** (1986), 820-821.

3. Walter Feit, Roger Lyndon, and Leonard L. Scott, A remark about permutations, *Journal of Combinatorial Theory* (**A**) **18** (1975), 234–235.

4. Lawrence Fialkow and Hector Salas, Data exchange and permutation length, *Mathematics Magazine* **65** (1992), 188–193.

George Mackiw
Loyola College
Contributed

PART 9
Other Topics

Introduction

The topics of linear algebra cannot be limited to those that fit into the previous eight topics. This section contains such items, which are none the less of value to instructors (and students) of linear algebra. They could fit into a variety of linear algebra courses depending on the emphasis of a particular course.

Positive Definite Matrices

Historically, positive definite matrices arise quite naturally in the study of n-ary quadratic forms and assume both theoretic and computational importance in a wide variety of applications. For example, they are employed in certain optimization algorithms in mathematical programming, in testing for the strict convexity of scalar-valued vector functions (here, positive definiteness of the Hessian provides a sufficiency check), and are of basic theoretic importance in construction of the various linear regression models. These are only a few of the specific applications which may be added to the abstract interest of such matrices. We now concentrate specifically on the properties of the matrices themselves.

DEFINITION: *An $n \times n$ real matrix A, where n is a positive integer, is called positive definite if $(x, Ax) = x^T Ax > 0$ for all nonzero column vectors x in Euclidean n-dimensional space.*

We shall designate the set of all such matrices (which forms a subset of all $n \times n$ matrices) as $\prod_n$.

Classically, it is customary to require also symmetry in the definition of positive definite, and we shall often concentrate on that proper subset of $\prod_n$ which consists of only the symmetric members of $\prod_n$. We shall designate this subset as $\sum_n$, but at times we shall allow ourself to consider the more general case.

It will become clear that the classical concentration on $\sum_n$ is convenient since it is much richer in algebraic properties, but also, from the standpoint of testing arbitrary matrices, it suffices to consider the theory of $\sum_n$. As is well known, any square matrix A can be written as the sum of a symmetric and a skew-symmetric matrix, $A = B + C$, where $B = (A + A^T)/2$ and $C = (A - A^T)/2$. We call B the *symmetric* and C the *skew-symmetric* part of A, in an unambiguous manner.

(1) REMARK. An $n \times n$ matrix A is positive definite if and only if the symmetric part of A is positive definite. (Actually, we show that any quadratic form is equivalent to a symmetric quadratic form.)

Proof. If C is the skew-symmetric part of A, then $(x, Cx) = (C^T x, x) = (-Cx, x) = -(Cx, x)$ implies $(x, Cx) = 0$. Therefore, $(x, Ax) = (x, (B + C)x) = (x, Bx)$, where B is the symmetric part of A. Thus, in a certain sense (e.g., that of testing for positive definiteness), it suffices to study the subset $\sum_n$. The possibility of generalizing some of the results to $\prod_n$ will also be discussed.

We first characterize positive definiteness $(\sum_n)$ in terms of a basic matrix invariant.

(2) THEOREM. *If A is $n \times n$ symmetric, then $A \in \sum_n$ if and only if all eigenvalues of A are positive.*

Proof. We first note that since A is symmetric, A may be diagonalized by some *orthogonal* matrix $B, B^T A B = D$ where $B^T = B^{-1}$ and D is diagonal with the necessarily real eigenvalues of A (Principal Axis Theorem, see e.g., Zelinsky, [3]). Let the eigenvalues of A be $\lambda_1, \ldots, \lambda_n$.

If all $\lambda_i > 0$, let $y = B^T x$ so that $x = By$. Then $x^T A x = (By)^T A(By) = y^T (B^T A B) y > 0$ if $x \neq 0$.

If A is positive definite, let v_i be an eigenvector corresponding to λ_i normalized so that $(v_i, v_i) = 1$. Then $0 < v_i^T A v_i = (\lambda_i v_i, v_i) = \lambda_i$ and all $\lambda_i > 0$.

(3) COROLLARY. *If $A \in \sum_n$, then $\det(A) > 0$.*

Proof: $\det(A) = \lambda_1 \cdots \lambda_n > 0$, by (2).

To facilitate the following comments, we let $S \subset \{1, 2, \cdots, n\}$, properly, be an index set. Then, let A_S, where A is an $n \times n$ matrix, be the matrix obtained from A by eliminating the rows and columns indicated by S, thus reducing the size of A.

(4) THEOREM. *If $A \in \prod_n$, then A_S is positive definite for any S. In particular, the diagonal elements of A are > 0.*

Proof: Let $x \neq 0$ be an n-vector with zeros as the components indicated by S and arbitrary components elsewhere. If x_S is the vector obtained from x by eliminating the (zero) components indicated by S, then $x_S^T A_S x_S = x^T A x > 0$. Since $x_S \neq 0$ is arbitrary, A_S is positive definite.

As a special case, we may let $S = \{1, \cdots, i - 1, i + 1, \cdots, n\}$ to show that the ith diagonal element of A is > 0.

We are now in a position to characterize positive definiteness in another manner which may be viewed as a test, the familiar determinant criteria. For this we employ the following abbreviations. If S is of the form $\{i + 1, i + 2, \cdots, n\}$, we denote A_S as A_i, i.e., A_i is the $i \times i$ matrix formed from the "intersection" of the first i rows and columns of A.

(5) THEOREM. *If A is $n \times n$ symmetric, then $A \in \sum_n$ if and only if $\det(A_i) > 0$ for $i = 1, \cdots, n$.*

Proof. If $A \in \sum_n$, then $A_i \in \sum_i$ by (4) and because A_i is symmetric. Therefore $\det(A_i) > 0$ by (3).

Unfortunately, there are not thoroughly pleasing proofs of the converse proposition, but the Inclusion Principle for eigenvalues (Franklin, [1]) will aid in the proof. We note that if $\alpha_1 \geq \cdots \geq \alpha_n$ are the n eigenvalues of A, and $\beta_1 \geq \cdots \geq \beta_{n-1}$ are the $n - 1$ eigenvalues of A_{n-1}, then $\alpha_1 \geq \beta_1 \geq \alpha_2 \geq \beta_2 \cdots \geq \alpha_{n-1} \geq \beta_{n-1} \geq \alpha_n$. Now, if $\det(A_i) > 0$ for $i = 1, \cdots, n$, we may inductively show $A \in \sum_n$. Since $\det(A_1) > 0, A_1 \in \sum_1$. If $A_k \in \sum_k$ for $k < n$, all eigenvalues of A_k are > 0, and thus by the Inclusion Principle, all eigenvalues of A_{k+1} are greater than 0, except perhaps the smallest. But, let $\alpha_1 \geq \cdots \geq \alpha_{k+1}$ be the eigenvalues of A_{k+1}; then $\alpha_{k+1} = \det(A_{k+1})/\alpha_1 \cdots \alpha_k$ as in the proof of (3), and α_{k+1}, the quotient of two positive reals, is positive. Thus, $A_{k+1} \in \sum_{k+1}$, and $A = A_n \in \sum_n$ by induction.

We choose to state here without full proof a result which applies in a much more general setting than $\prod_n$, but which is important in applications of positive definite matrices.

(6) THEOREM. *If $A \in \prod_n$, then A has a unique factorization $A = LR$ into triangular*

factors where L is lower triangular with 1's on the diagonal and R is upper triangular with nonzero diagonal elements.

Comments. This is a consequence of the more general theorem (e.g., see [1], p. 204) that A has such a factorization if and only if $\det(A_i) \neq 0$ for $i = 1, \cdots, n$. Since $A \in \prod_n$, each $A_i \in \prod_i$ by (4). Thus $\det(A_i) \neq 0$, for otherwise A_i would be singular and not positive definite. Since L and R are obviously invertible (with easily computed inverses), this theorem is helpful, for instance, in exhibiting solutions x to $Ax = y$.

We now have three rather concrete characterizations of $\sum_n$. With little difficulty we may add a fourth which is not quite so concrete.

(7) LEMMA. *IF $A \in \sum_n$, then there exists an invertible matrix P such that $P^T A P = I$, the identity matrix. Also $PP^T = A^{-1}$ which exists.*

Proof: We know there exists an orthogonal B, such that $B^T A B = C$, where C is diagonal with the eigenvalues of A. If we let $D = +C^{-1/2}$, then D is well defined. Now let $P = BD$. Then, P is invertible since B and D are, and $P^T A P = D^T B^T A B D = DCD = CD^2 = CC^{-1} = I$. Also, $P^T A P P^T = P^T$ and $APP^T = I$ since P^T is invertible. Therefore, $A^{-1} = PP^T$.

As an aside, (7) points out that any symmetric positive definite quadratic form $x^T Ax$ (in fact, any positive definite quadratic form by (1)), is equal to (y, y) under a suitable change of coordinates.

(8) THEOREM. *$A \in \sum_n$ if and only if $\exists Q$ invertible such that $A = Q^T Q$.*

Proof. If $A = Q^T Q$ and $y = Qx$, then $x^T A x = x^T Q^T Q x = y^T y = (y, y) > 0$ if $x \neq 0$ since Q is assumed invertible.

If $A \in \sum_n$, then $A = (P^T)^{-1} P^{-1}$, with P invertible by (7). Let $Q = P^{-1}$; then $Q^T = (P^{-1})^T = (P^T)^{-1}$.

We now mention a few less fundamental but still important results which deal with the positive definiteness of some functions of positive definite matrices.

(9) THEOREM. *The matrix A belongs to $\sum_n$ if and only if $B^T A B$ belongs to $\sum_m$ for each $n \times m$ matrix B such that $By = 0$ implies $y = 0$.*

Proof. If the condition is satisfied, let $m = n$ and choose $B = I$. Then $I^T A I = A \in \sum_n$.

Conversely, let $A \in \sum_n$ and suppose B satisfies the conditions of the theorem. Then $y^T(B^T A B)y = (By)^T A(By) > 0$ unless $By = 0$, that is, $y = 0$. Hence, $B^T A B$ is symmetric and in $\sum_m$.

(10) THEOREM. *If $A \in \sum_n$ then*

(a) *$cA \in \sum_n$ for $c > 0$ any real scalar;*
(b) *$(A + B) \in \sum_n$ if $B \in \sum_n$;*
(c) *$A^m \in \sum_n$ for m any integer;*
(d) *an $A^{1/p}$ exists $\in \sum_n$ for p a positive integer (by $A^{1/p}$ we mean a matrix B such that $B^p = A$);*
(e) *an A^r exists $\in \sum_n$ for r any rational number.*

Proof.

(a) cA is symmetric, and $x^T(cA)x = c(x^T Ax) > 0$ if $c > 0$ and $x \neq 0$.

(b) $x^T(A + B)x = x^T(Ax + Bx) = x^T Ax + x^T Bx > 0$ if $x \neq 0$. Also, $A + B = (A + B)^T$ is symmetric.

(c) $(A^m)^T = (A^T)^m = A^m$ is symmetric. If $m = 0, A^m = I \in \Sigma_n$. If $m > 0$ and $C^T AC = D$ is an orthogonal diagonalization of A, then $A = CDC^T$ and $A^m = (CDC^T)(CDC^T)\cdots(CDC^T) = CD^m C^T$. Thus A^m is diagonalizable to D^m, and, therefore, has all eigenvalues > 0. Therefore, $A^m \in \Sigma_n$. If $m = -1$, then A^m exists in Σ_n by (7) and (8). If $m < -1, -m > 0, A^m = (A^{-1})^{-m}$, and $A^m \in \Sigma_n$.

(d) As before, we may write $A = CDC^T$, where C is orthogonal and D diagonal with positive diagonal elements. Define $A^{1/p} = CD^{1/p}C^T$, where $D^{1/p}$ is diagonal with diagonal elements the positive real pth roots of the diagonal elements of D. Then $(A^{1/p})^p = A$, $A^{1/p}$ is symmetric and has n positive real eigenvalues and is thus positive definite.

(e) Follows from (c) and (d).

The theorem, (10), might give us hope that a nontrivial algebraic structure might be imposed on Σ_n (or, perhaps, Π_n) so that it could be characterized as one of the more familiar algebraic objects. Unfortunately, however, this does not seem to be the case.

We may not employ an additive group structure since inverses do not exist in Σ_n. If A is positive definite, not only is $-A$ not positive definite, but it has essentially the opposite properties of A (such a matrix is called *negative definite*). Also, there is no additive identity in Σ_n or Π_n. Of course, both Σ_n and Π_n are closed and associative under addition.

The situation is interestingly different but only a little more well behaved if we attempt a multiplicative group structure on Σ_n. We have the identity matrix $I \in \Sigma_n$ and inverses exist in Σ_n by (10). Also, the multiplication is associative. But Σ_n is not closed under matrix multiplication. Not only is the product of two members of Σ_n not symmetric if they do not commute, but it may not even be positive definite as the following example in Σ_2 shows. Let

$$A = \begin{pmatrix} 1 & 3 \\ 3 & 10 \end{pmatrix} \qquad \text{and} \qquad B = \begin{pmatrix} 1 & -3 \\ -3 & 10 \end{pmatrix}.$$

Both A and B are in Σ_2 by (5). However,

$$AB = \begin{pmatrix} -8 & 27 \\ -27 & 91 \end{pmatrix} \notin \Pi_2$$

by (4). In fact, AB is neither positive nor negative definite.

About all we can say, then, is that Π_n forms a semigroup under matrix addition. Since the other common algebraic structures are fundamentally more complex than the group, they are also precluded.

Some generalizations may be made without great difficulty on the proofs exhibited thus far. For instance, in the "only if" part of (2) our hypothesis is unnecessarily strong. If A is

$n \times n$, it need only have n real eigenvalues. The proof is easily seen to depend only on the positive definiteness of A and the existence of n real eigenvalues with associated eigenvectors. Thus we may validly formulate the amending statement:

(2.1) THEOREM. *If A is $n \times n$ and has n real eigenvalues, then $A \in \prod_n$ implies all the eigenvalues of A are positive.*

An entirely analogous statement may be made about the corollary, (3).

We may not, however, drop the requirement that A is real diagonalizable from (2.1) since a matrix may be positive definite and have complex characteristic roots as the following example shows. Let

$$A = \begin{pmatrix} 2 & 1 \\ -1 & 2 \end{pmatrix}.$$

Then $A \in \prod_2$ by (1) since the symmetric part of A is

$$\begin{pmatrix} 2 & 0 \\ 0 & 2 \end{pmatrix}.$$

However, the characteristic polynomial of A is $\lambda^2 - 4\lambda + 5$ with roots $2 \pm i$. It should be clear, though, that under no circumstances may a member of $\prod_n$ have a nonpositive real eigenvalue; for then we could choose a nonzero eigenvector to violate the definition of $\prod_n$. Thus, to generalize slightly on (7), a member of $\prod_n$ must always have nonzero determinant and be invertible.

In (10) parts (a) and (b) may clearly be generalized to all of $\prod_n$, but part (c) is not always valid in $\prod_n$. For instance, if

$$A = \begin{pmatrix} 1 & -4 \\ 2 & 10 \end{pmatrix},$$

$A \in \prod_2$ by (1) and (5). But

$$A^2 = \begin{pmatrix} -7 & -44 \\ 22 & 92 \end{pmatrix}$$

which is not positive definite by (4). Also, this example should caution us to note that a matrix may have all its eigenvalues real and positive but not belong to any $\prod_n$. A^2 has eigenvalues $\{81, 4\}$.

If we allow complex-entried matrices and then define $\sum_n^H$ to be the *Hermitian* matrices which are positive definite, it is clear also that our results concerning $\sum_n$ may be modified to remain valid in $\sum_n^H$. In addition we should note that our notion of $\prod_n$ has no consistent analog in the complex-entried matrices. For instance, a matrix $A \in \prod_n$ with respect to real vectors may not even have $x^* A x$ real when complex vectors are allowed ($*$ means "conjugate transpose").

We have thus far formulated a theory of positive definite matrices. It is clear that we may analogously define another (disjoint) set of matrices by replacing ">" with "<" in the definition of positive definite. Such matrices are usually termed *negative definite*, and, suggestively, we might designate this set as $-\prod_n$ since $A \in \prod_n$ if and only if $-A \in -\prod_n$. This, of course, is the key to the development of a theory of negative definite matrices which would proceed analogously (allowing for the peculiarities of negative numbers).

Positive (negative) *semi-definite* matrices may be defined by allowing the possibility of equality in the definition of $\prod_n$ (or $-\prod_n$). Their theory proceeds similarly, but modified by allowance for 0 eigenvalues.

In the positive definite case, we have succeeded in establishing four characterizations through theorems: by eigenvalues ((2)); by determinants ((5)); by triangular decomposition ((6)); and by $Q^T Q$ decomposition ((8)). This, plus the additional properties commented on, is largely sufficient to both mathematically describe and usefully apply positive definite matrices.

References

1. J. N. Franklin, *Matrix Theory*, Prentice Hall, Englewood Cliffs, N.J., 1968.
2. A.S. Goldberger, *Econometric Theory*, Wiley, New York, 1966.
3. D. Zelinsky, *A First Course in Linear Algebra*, Academic Press, New York, 1968.

C.R. Johnson
Northwestern University
American Mathematical Monthly **77** (1970), 259–264

Quaternions

In this section we study the Hamiltonian quaternions and examine their relationship to a subset of $M_4(\mathbb{R})$, the set of real 4×4 matrices. The set of quaternions is an algebraic system that was first described by the Irish mathematician Sir William Rowan Hamilton (1805–1865). Hamilton, who at the age of five could read Latin, Greek, and Hebrew, studied the works of Clairaut and Laplace as a boy. Young Willliam's mathematical career began by finding a mistake in Laplace's *Mécanique céleste*. While still an undergraduate at Trinity College, Dublin, he was appointed Professor of Astronomy there. He discovered the quaternions at the age of 38 while on a walk with his wife. Hamilton later described the discovery with these words, ". . . I then and there felt the galvanic circuit of thought closed, and the sparks which flew from it were the fundamental equations between $\mathbf{I}, \mathbf{J}, \mathbf{K} \ldots$" [1].

Hamilton believed that his discovery was as important as calculus and that it would become an indispensable tool in mathematical physics. Even though the quaternions were applied in optics and mechanics, the work did not enjoy wide spread acceptance in the physics community. However, the quaternions proved to be enormously important in algebra. Additionally, the quaternions proved to be a very important example—one that shares many algebraic properties of the real numbers but has a non-commutative multiplication.

The set of quaternions, denoted by $\mathbb{H}$, is

$$\left\{ a + bi + cj + dk \mid a, b, c, \text{ and } d \text{ are real numbers}, \ i^2 = j^2 = k^2 = -1 \right\}.$$

Addition of two quaternions, $a + bi + cj + dk$ and $e + fi + gj + hk$, is defined by

$$(a + bi + cj + dk) + (e + fi + gj + hk) = (a + e) + (b + f)i + (c + g)j + (d + h)k.$$

Multiplication of two quaternions, $a + bi + cj + dk$ and $e + fi + gj + hk$, is defined by

$$(a + bi + cj + dk)(e + fi + gj + hk) = (ae - bf - cg - dh) + (af + be + ch - dg)i + (ag - bh + ce + df)j + (ah + bg - cf + de)k.$$

It is easily shown that the units i, j, and k obey the following rules of multiplication:

$$ij = k, \ jk = i, \ ki = j, \ ik = -j, \ kj = -i, \ ji = -k$$

The quaternions are an extension of the real numbers (take $b = c = d = 0$ in the definition of $\mathbb{H}$) and the complex numbers (take $c = d = 0$ in the definition of $\mathbb{H}$). $\mathbb{H}$ and $\mathbb{R}$, enjoy many of the same properties. Some of these are associativity of addition and multiplication; commutativity of addition; additive and multiplicative identities; additive inverses; and multiplicative inverses for non-zero elements. However, multiplication in $\mathbb{H}$ is clearly not commutative as the above calculations with i, j, and k demonstrate. Another anomaly that occurs is that an nth degree polynomial equation in $\mathbb{H}$ can have more than n solutions in $\mathbb{H}$. Indeed it is possible to have a polynomial equation in $\mathbb{H}$ which has infinitely many solutions. Such a thing could not occur in $\mathbb{R}$ or $\mathbb{C}$.

The formula for quaternion addition is quite easy to implement. To the contrary, the formula for quaternion multiplication is very tedious. We shall see that quaternions can be represented by 4×4 real matrices and this representation will facilitate calculations. To do this we state a number of results which we hope some of you will prove.

$\mathbb{H}$ is a real 4-dimensional vector space having $\{1, i, j, k\}$ as a basis. Consider $M_4(\mathbb{R})$, the real vector space of 4×4 matrices and define $L : \mathbb{H} \to M_4(\mathbb{R})$ by

$$L(a + bi + cj + dk) = \begin{bmatrix} a & b & c & d \\ -b & a & -d & c \\ -c & d & a & -b \\ -d & -c & b & a \end{bmatrix}.$$

L is a one-to-one linear transformation that also preserves multiplication, i.e.,

$$\begin{aligned} L((a + bi + cj + dk)(e + fi + gj + hk)) = \\ L(a + bi + cj + dk)L(e + fi + gj + hk) \end{aligned}.$$

The range of $L, Rng(L)$, is a 4-dimensional subspace of $M_4(\mathbb{R})$. Indeed, $\mathbb{H}$ and $Rng(L)$ are "isomorphhic" as vector spaces—that is to say, from a vector space point of view, $\mathbb{H}$ and $Rng(L)$ are indistinguishable.

If $z = a + bi + cj + dk$ is a quaternion, then the conjugate of z, denoted by $\bar{z}$, is defined by

$$\bar{z} = a - bi - cj - dk.$$

The norm denoted by $N(z)$, is defined by

$$N(z) = z\bar{z}.$$

It is clear that

$$\bar{\bar{z}} = z.$$

Note that

$$L(\overline{a + bi + cj + dk}) = \begin{bmatrix} a & b & c & d \\ -b & a & -d & c \\ -c & d & a & -b \\ -d & -c & b & a \end{bmatrix}^T.$$

Here the superscript "T" denotes the transpose operator.

The following exercises are quite appropriate from a pedagogical point of view, but may be outside of the scope of a sophomore linear algebra course. Hopefully this caveat will not deter you from attempting these exercises.

* Prove that $\mathbb{H}$ is a real vector space of dimension four.

* Prove that $\mathbb{H}$ is a complex vector space of dimension two.

* Prove that L is a linear transformation.

* Prove that L preserves multiplication.

For a further discussion of quaternions, see [2].

Exercise Set

Use the relationship between $\mathbb{H}$ and $Rng(L)$ to give matric solutions to the following problems. Express your answer in the form $a + bi + cj + dk$ when appropriate.

Let $a = 5 + 4i + 2j - k, b = 6 + 3i - 9j - 4k, c = 2 + 2i + 3j + 6k$, and $d = 1 + 7i - j + k$. In exercises 1–5 perform the indicated calculation.

1.   a. $a + b$
        b. $c - d$
2.   a. $4a - 6b$
        b. $-11c + 10d$
3.   a. a^{-1}
        b. c^{-1}
4. $dN(d)^{-1}$
5. $N\left((dN(d)^{-1})\right)$
6.   Compute $d\bar{d}$ and $\bar{d}d$. Compare and comment on the results.
7.   Compute $\overline{bc}$ and $\bar{b}\bar{c}$. Compare and comment on the results.
8.   Computer $\overline{b - a}$ and $\bar{b} - \bar{a}$. Compare and comment on the results.
9.   Compute ab^{-1} and ba^{-1}. Compare and comment on the results.
10. Compute $\bar{a}(a\bar{a})^{-1}$ and a^{-1} (see 3a). Compare and comment on the results.
11. Compute $a^{-1}da$ and $a^{-1}d^{-1}a$. Compare and comment on the results.
12. Compute $\bar{a}^{-1}$ and $\overline{a^{-1}}$ Compare and comment on the results.

Solve the following equations for x. Check the accuracy of your answer using an equivalent matric equation and your favorite computer algebra system.

13. $\overline{ax} - \bar{a} = d^{-1}\bar{b}$
14. $-\left(xb^{-1} - \bar{b}\right) = ((\bar{a} - c)\, d^{-1})^{-1}$

In the following exercises let $x = 2i/3 - j/3 + 2k/3, y = 2i/7 - 3j/7 - 6k/7, z = -4i/9 + j/9 - 8k/9$.

16. Compute each of the following.
    a. x^2
    b. y^2
    c. z^2
17. What is the algebraic significance of your calculations in Exercise 16. [*Hint*: Consider the equation $w^2 + 1 = 0$.]
18. Can you see a pattern for the coefficients of x, y, and z that relates to geometric ideas? [*Hint*: Consider the sphere with center at the origin and radius one.]
19. In light of your response to Exercise 18 make a conjecture about the solutions to $w^2 + 1 = 0$. What justification can you give for your conjecture?

References

1. M. Kline, *Mathematical Thought from Ancient to Modern Times*, Oxford Press, New York, 1972.

2. G. Birkhoff and S. MacLane, *A Survey of Modern Algebra*, Revised Edition, The Macmillan Company, New York, 1953.

Robert S. Smith
Miami University
Contributed

Editors' Note: The interested instructor may wish to see other elementary discussions of quaternions. We recommend the following.

Hamilton's Discovery of the Quaternions, *Math Magazine* **45**, Number 5, 1976, p. 227.

Hamilton, Rodrigues, and the Quaternion Scandal, *Math Magazine* **62**, Number 5, 1989, p. 291.

UMAP modules 313, 652, 722.

Bionomial Matrices

One aim of this paper is to interconnect bionomial coefficients with matrix algebra and thereby obtain "new wine in new bottles." Another purpose is to exhibit a non-trivial class of matrices whose tth power can be obtained at sight, for all rational numbers t.

Consider the following table:

Table 1

	col. 0	col. 1				col. j
row 0	1					
row 1	x	y				
	x^2	$2xy$	y^2			
row r	x^r	$\binom{r}{1}x^{r-1}y$	$\binom{r}{2}x^{r-2}y^2$	$\binom{r}{3}x^{r-3}y^3$		$\binom{r}{j}x^{r-j}y^j$

Observe that the top row in the above table is labeled row 0 and the leftmost column is labeled column 0. The entries in the cells of row r are the successive terms in the binomial expansion of $(x+y)^r$; the entry in row r, column j is $\binom{r}{j}x^{r-j}y^j$. By definition, a Bionomial Matrix of order k is obtained by taking the rows and columns of the above table, consecutively, from 0 through $(k-1)$. We denote such a matrix by $B_k(x, y)$. (Whenever the order is of no importance, or clear from the context, we drop the subscript k.)

We use the symbol P_k to denote the special case where $x = y = 1$, i.e.

$$P_k = B_k(1, 1).$$

Observe that P_k corresponds to the first k rows and columns of Pascal's triangle. Note further that the jth column of $B(x, y)$ begins with j zeros, followed by the successive terms in the "negative bionomial expansion" of $y^j(1-x)^{-j-1}$.

Next, consider Table 2 below. This is a table of coefficients, a well-known symmetric form of Pascal's triangle which we call the S-table.

273

Table 2

$i\backslash j$	Column Number						
	0	1	2	3	4		r
0	1	1	1	1	1		$\binom{r}{r}$
1	1	2	3	4	5		$\binom{r+1}{r}$
2	1	3	6	10	15		$\binom{r+2}{r}$
3	1	4	10	20	35		$\binom{r+3}{r}$
4	1	5	15	35	70		$\binom{r+4}{4}$
r	$\binom{r}{0}$	$\binom{r+1}{1}$	$\binom{r+2}{2}$	$\binom{r+3}{3}$	$\binom{r+4}{4}$		$\binom{2r}{r}$

We define the S_k matrix as consisting of the first k rows and columns of the S-table. Observe that the entry in row i, column j of the S-table is equal to the entry in row $(i+j)$, column j of Pascal's triangle, namely $\binom{i+j}{j}$.

As an illustration of the relationship of these matrices, we have

Theorem 1. $P_k P_k^T = S_k$, for each integer $k > 0$. (P_k^T denotes the transpose of P_k).

To illustrate this result, consider

$$P_3 P_3^T = \begin{pmatrix} 1 & 0 & 0 \\ 1 & 1 & 0 \\ 1 & 2 & 1 \end{pmatrix} \begin{pmatrix} 1 & 1 & 1 \\ 0 & 1 & 2 \\ 0 & 0 & 1 \end{pmatrix} = \begin{pmatrix} 1 & 1 & 1 \\ 1 & 2 & 3 \\ 1 & 3 & 6 \end{pmatrix} = S_3.$$

We point out that $P_3 P_3^T$ does not equal $P_3^T P_3$.

By the definition of matrix multiplication, Theorem 1 follows from the combinatorial identity

$$\binom{i}{0}\binom{j}{0} + \binom{i}{1}\binom{j}{1} + \binom{i}{2}\binom{j}{2} + \cdots + \binom{i}{i}\binom{j}{i} = \binom{i+j}{j}, \text{ for } i \le j. \tag{1}$$

This identity follows from examination of the polynomial identity

$$(1+x)^i(1+x^{-1})^j = (1+x)^i(1+x)^j x^{-j} = x^{-j}(1+x)^{i+j} \tag{2}$$

If we equate the constant term in the left-hand side of Equation (2) with the constant term in the right-hand side, then Equation (1) follows immediately.

In relation to general matrix theory, Theorem 1 is incidental. (Observe, however, certain interesting exercises, e.g. $\det(S_n) = 1$.)

The following results seem to be of more general interest:

Theorem 2. *Binomial matrix multiplication satisfies:*

$$B_k(x,y)B_k(w,z) = B_k(x+yw, yz) \text{ for each } k = 1, 2, \ldots .$$

Observe that the entry in row r, column j of the above matrix product is the dot product of vectors (a) and (b) below:

(a) $\binom{r}{0}x^r, \quad \binom{r}{1}x^{r-1}y, \ldots, \quad \binom{r}{j}x^{r-j}y^j, \quad \binom{r}{j+1}x^{r-j-1}y^{j+1}, \ldots, \quad \binom{r}{r}y^r, 0, \ldots 0.$

(b) $\binom{0}{j}, \quad \binom{1}{j}, \ldots, \quad \binom{j}{j}z^j, \quad \binom{j+1}{j}wz^j, \ldots, \quad \binom{r}{j}w^{r-j}z^j, \ldots .$

Since $\binom{k}{j} = 0$ for all $k < j$, this dot product equals

$$\binom{r}{j}\binom{j}{j}x^{r-1}y^j z^j + \binom{r}{j+1}\binom{j+1}{j}x^{r-j-1}y^{j+1}wz^j + \cdots + \binom{r}{r}\binom{r}{j}y^r w^{r-j}z^j. \qquad (3)$$

It is simple to verify the identity

$$\binom{r}{j+t}\binom{j+t}{j} = \binom{r}{j}\binom{r-j}{t}.$$

Accordingly, we can rewrite Equation (3) as

$$\binom{r}{j}y^j z^j \left[\binom{r-j}{0}x^{r-j} + \binom{r-j}{1}x^{r-j-1}(yw) + \cdots + \binom{r-j}{r-j}(yw)^{r-j} \right]$$

$$= \binom{r}{j}(yz)^j \left[x + yw\right]^{r-j},$$

which establishes Theorem 2.

Remark: As an example of Theorem 2 let

$$B(x,y) = \begin{pmatrix} 1 & 0 \\ x & y \end{pmatrix} \text{ and } B(w,z) = \begin{pmatrix} 1 & 0 \\ w & z \end{pmatrix}.$$

By direct multiplication,

$$B(x,y)B(w,z) = \begin{pmatrix} 1 & 0 \\ x+yw & yz \end{pmatrix}.$$

Corollary 2.1. *If $y \neq 0$, then $B(x, y)$ is invertible and*

$$[B(x, y)]^{-1} = B(-xy^{-1}, y^{-1}).$$

Proof. By Theorem 2, we have

$$B(x, y)B(-xy^{-1}, y^{-1}) = B(0, 1) = I_k,$$

(the $k \times k$ identity).

In particular,

$$B(x, 1)^{-1} = B(-x, 1) \text{ and}$$
$$B(1, y)^{-1} = B(-y^{-1}, y^{-1}).$$

Corollary 2.2. $B(x, y)$ *and* $B(w, z)$ *commute if and only if* $x(1 - z) = w(1 - y)$.

This follows from comparing

$$B(x, y)B(w, z) = B(x + yw, yz)$$

and

$$B(w, z)B(x, y) = B(w + zx, zy).$$

Whenever $x + y = 1$, we call $B(x, y)$ *generalized stochastic*. A generalized stochastic matrix will be said to be *strictly stochastic*, or simply *stochastic* if x and y are non-negative.

The following corollary is an immediate consequence of Theorem 2:

Corollary 2.3. *Any two generalized stochastic binomial matrices commute.*

Since, by L'Hospital's rule

$$\lim_{y \to 1} \left(\frac{1 - y^t}{1 - y} \right) = t,$$

we use the *convention* that $(1 - y^t / 1 - y) = t$ whenever $y = 1$.

Theorem 3. $[B(x, y)]^n = B(x\,[(1 - y^n / 1 - y)], y^n)$, *for any integer* $n > 0$.

From Theorem 2 it follows that

$$[B(x, y)]^2 = B(x[1 + y], y^2),$$

which shows that Theorem 3 holds when $n = 2$ (in either case, $y \neq 1$ or $y = 1$). Assume the theorem true for all integers less than or equal to $(n - 1)$.

(a) If $y \neq 1$ then

$$[B(x, y)]^n = [B(x, y)]^{n-1} \cdot B(x, y) = B\left(x\left[\frac{1 - y^{n-1}}{1 - y}\right], y^{n-1} \right) \cdot B(x, y)$$

$$= B\left(x\left\{ \left[\frac{1 - y^{n-1}}{1 - y}\right] + y^{n-1} \right\}, y^n \right)$$

$$= B\left(x\left[\frac{1 - y^n}{1 - y}\right], y^n \right).$$

(b) If $y = 1$, then (as remarked above, or directly from Theorem 2)

$$B(x, 1)B(x, 1) = B(2x, 1).$$

By induction, it follows that

$$B(x, 1)^n = B(nx, 1),$$

which, when using our convention, is the statement of Theorem 3 for $y = 1$.

Observe that Theorem 3 is also true for $n = 0$ (unless $y = 0$), since both sides reduce to I_k.

We now extend Theorem 3 to be valid for any integer n:

Theorem 4. $B(x, y)^n = B(x[(1-y)^n/(1-y)], y^n)$ *for any integer n, provided $y \neq 0$.*

We have already proved this for n positive or zero. Now let n be positive, so that $m = -n$ is negative. Then, for $y \neq 0$,

$$B(x, y)^{-n} = [B(x, y)^{-1}]^n$$

$$= B(-xy^{-1}, y^{-1})^n, \text{ by Corollary 2.1,}$$

$$= B\left[-xy^{-1}\left(\frac{1-y^{-n}}{1-y^{-1}}\right), y^{-n}\right], \text{ by Theorem 3,}$$

$$= B\left[x\left(\frac{1-y^{-n}}{1-y}\right), y^{-n}\right];$$

that is,

$$B(x, y)^m = B\left(x\left[\frac{1-y^m}{1.-y}\right], y^m\right).$$

Corollary 4.1. $P^n = B(1, 1)^n = B(n, 1)$ *for all integers n.*

To the best of our knowledge, Corollary 4.1 was first proved in the special case $n = -1$ by John Riordan [1].

We define for any rational t

$$B(x, y)^t = B\left(x\left[\frac{1-y^t}{1-y}\right], y^t\right).$$

This definition will only make sense if y^t is well-defined. Furthermore, if $y = 1$, by convention, $(1-y^t)/(1-y) = t$.

In particular, for any integers $m, n \neq 0$ and $y \neq 0$,

$$B(x, y)^{m/n} = B\left(x\left[\frac{1-y^{m/n}}{1-y}\right], y^{m/n}\right),$$

provided that $y^{m/n}$ is well-defined. This makes sense since, by Theorem 4,

$$B\left(x\left[\frac{1-y^{1/n}}{1-y}\right],y^{1/n}\right)^n = B\left(\left[x\frac{1-y^{1/n}}{1-y}\right]\left[\frac{1-(y^{1/n})^n}{1-y^{1/n}}\right],y^{(1/n)^n}\right)$$
$$= B(x,y);$$

and

$$\left[B(x,y)^{1/n}\right]^{-m} = B\left(x\frac{1-y^{1/n}}{1-y},y^{1/n}\right)^m$$
$$= B\left(x\frac{1-y^{1/n}}{1-y}\frac{1-(y^{1/n})^m}{1-y^{1/n}},(y^{1/n})^m\right)$$
$$= B\left(x\frac{1-y^{m/n}}{1-y},y^{m/n}\right)$$
$$= B(x,y)^{m/n}.$$

It is straightforward to show that the definition

$$[B(x,y)]^t = B\left(x\left[\frac{1-y^t}{1-y}\right],y^t\right)$$

will surely give a unique tth power provided that

(i) y^t is defined to be the unique, rational positive tth power of y.

(ii) $(1-y^t)/(1-y) = t$ whenever $y = 1$.

(iii) We limit the class of admissible tth powers to binomial matrices whose entries are real numbers.

In particular, if B_k is a stochastic matrix we can obtain a well-defined unique tth power. Such a concept is quite useful—suppose, for example, that readings on a Markov chain are taken every three months. Then one possible answer to the question of "what went on monthly" is to take a unique cube-root of the observed matrix. For an example of a Markov chain which is also a binomial matrix, consider the Estes' Learning Model when the experimenter always makes the same choice [2].

To illustrate, consider

$$B_3 = \begin{bmatrix} 1 & 0 & 0 \\ x & y & 0 \\ x^2 & 2xy & y^2 \end{bmatrix}.$$

We define $B_3^{1/2}$ by referring to the definition of $B(x,y)^t$ above. Accordingly,

$$
B_3^{1/2} = \begin{bmatrix} 1 & 0 & 0 \\ x\left(\dfrac{1-y^{1/2}}{1-y}\right) & y^{1/2} & 0 \\ x^2\left(\dfrac{1-y^{1/2}}{1-y}\right) & 2x\left(\dfrac{1-y^{1/2}}{1-y}\right)^{y^{1/2}} & y \end{bmatrix}.
$$

To illustrate further, suppose that $x = y = \frac{1}{2}$. Then $y^{1/2} \doteq 0.707$ and $(y^{1/2}-1)/(y-1) \doteq 0.586$; then

$$
B_3\left(\frac{1}{2},\frac{1}{2}\right) = \begin{bmatrix} 1 & 0 & 0 \\ \frac{1}{2} & \frac{1}{2} & 0 \\ \frac{1}{4} & \frac{1}{2} & \frac{1}{4} \end{bmatrix},
$$

whereas

$$
\left[B_3\left(\frac{1}{2},\frac{1}{2}\right)\right]^{\frac{1}{2}} = \begin{bmatrix} 1 & 0 & 0 \\ 0.293 & 0.707 & 0 \\ 0.086 & 0.414 & 0.500 \end{bmatrix}.
$$

We terminate with a description of the eigenvalues and eigenvectors of $B(x,y)$.

Theorem 5. *Given $B_k(x,y)$ where $y \neq 0, y \neq 1$. Then the eigenvalues of $B_k(x,y)$ are distinct, viz., $y^0 = 1, y, y^2, \dots, y^{n-1}$, and the corresponding eigenvectors are the columns of $B_k(x/(1-y),1)$, in order.*

Proof. Consider the matrix product

$$
B_k(x,y)B_k\left(\frac{x}{1-y},1\right) = B_k\left(x+\frac{yx}{1-y},y\right) = B_k\left(\frac{x}{1-y},y\right).
$$

This proves our contention, column by column, directly from the definition of eigenvector and eigenvalue.

Corollary 5.1. *If $B_k(x,y)$ is general stochastic, i.e. ,if $x+y=1$, then its eigenvectors are the columns of $B_k(1,1) = P_k$.*

Corollary 5.2.

$$
B_k(x,y) = B_k\left(\frac{x}{1-y},1\right) \cdot Y \cdot B_k\left(\frac{-x}{1-y},1\right),
$$

where Y is the diagonal matrix of eigenvalues of $B_k(x,y)$ and $y \neq 0, y \neq 1$.

Note that if we are to attempt to define the tth power of a binomial matrix by first passing to its diagonalization as in Corollary 5.2 and then taking the tth roots of the diagonal elements in the diagonal matrix Y above, we would come to the conditions and definition given earlier, i.e.,

$$[B(x,y)]^t = B\left(x\left[\frac{1-y^t}{1-y}\right], y^t \right).$$

This relation follows from direct multiplication of the matrices associated with the right-hand side of Corollary 5.2, Y replaced by Y^t, where Y^t is deduced from Y by replacing the scalar y by the scalar y^t.

The main purpose of this paper has been expressed in the first paragraph. An incidental purpose, of course, is the hope that some of the deeper results of matrix theory can influence the study of combinatorial equalities and inequalities—for example, can we obtain combinatorial inequalities by applying the theory of eigenvalues to the symmetric matrix S_k of Theorem 1? Furthermore, it is of interest to note that the matrices $B_k(x,y)$ and S_k occur unnamed in numerous contexts. For such an example, we refer the reader to any proof of the *Inclusion-Exclusion Principle.*

The author wishes to express appreciation to those persons who have helped him clarify this paper—in particular, thanks to Professors Norman Schaumberger, James Slifker, and Warren Page.

References

1. J. Riordan, Inverse Relations and Combinatorial Identities, *Amer. Math. Monthly* **71**, Number 5, (1964).
2. Kemeny, Snell and Thompson, *Introduction to Finite Mathematics, 2nd ed.* , Prentice-Hall, Englewood Cliffs, New Jersey, 1966, pp. 415–423.
3. C.L. Liu, *Introduction to Combinatorial Mathematics, Chapter IV*, McGraw-Hill, New York, 1968.

Jay Strum
New York University
College Mathematics Journal **8** (1977), 260–266

A Combinatorial Theorem on Circulant Matrices

Introductory treatments frequently say that matrices are just rectangular arrays of numbers, but, of course, the real intent is to develop their algebraic properties. Here, we go no further than matrices as formal or symbolic objects. Half the time, the elements will be *colors* rather than numbers. Matrices have a capacity for storing combinatorial information, and we want to exploit that. The following examples extend some classical problems, and provide motivation.

1. Problem of n queens (Gauss, ca. 1850). Can n mutually invulnerable queens be placed on an $n \times n$ chessboard? Surprisingly, it seems that the answer—affirmative for $n \geq 4$—has been known only since 1969 [1], [2]. We give a simple new proof of this fact, below. Here, however, our interest is in other surfaces with chessboard-like properties: the wrapped-around cylindrical or toroidal boards. Whether on the $8 \times 8, 9 \times 9$, or 10×10 versions, the enhanced diagonals can make trying to place immune queens a real exercise in futility.

When can n invulnerable queens be placed on $n \times n$ toroidal chessboard?

2. Ring dancers. N husbands have joined hands in a large circle, facing inward, while their N wives have linked arms in an inner circle, facing outward. The two groups are dancing in opposite directions. Now, freeze the action. Picking a husband at random, we expect that the person facing him is not his wife. On the other hand, it isn't unreasonable to suspect that, somewhere on the circle, two spouses are facing each other.

Is it always possible to freeze the action at an instant when no spouses are facing each other?

3. N pigeons, N holes. The pigeonhole principle says that it takes at least $N + 1$ pigeons, flying into N pigeonholes, to guarantee that one of the pigeonholes will contain several pigeons. Here is a classic application [3],[4],[5],[6]:

Suppose that a disk has been divided into $n = 2k$ congruent sectors with k sectors colored blue and k colored red. Let the same thing be done to a smaller disk. The problem is to show that the smaller disk can always be positioned concentrically within the larger so that at least half the sectors have matching color. We comment on the solution later. For our nonstandard application, which removes the critical $N + 1$st pigeon (see below), let n be arbitrary and, instead of using colors, label the sectors of each disk with the numbers

$1, 2, \ldots, n$ in any order. We can place the disks concentrically and talk about the sum of the integer labels in a sector.

Can the disks be positioned so that at least two sector sums are equal?

4. A beautiful deception. To perpetrate this admitted flim-flam, one needs an accomplice seated anonymously in the audience. Participants are asked to draw a clockface with the numbers $1, 2, \ldots, 12$ arranged in any order they please. A persuasive MC will stress the secrecy of each individual's arrangement, that there are 11! or nearly 40 million possibilities, etc. This being done, everyone is asked to note the largest sum given by three consecutive numbers on their permuted clockface. This becomes a player's *secret number*. Now the bidding starts. A person is to bid any number if they are convinced that no one else in the audience can undercut them with their secret number. For example, I would bid 27 if I felt certain that no one else had a secret number of 26 or less. The winner is either the last bidder or the *first* one who undercuts another bidder. It's a game of nerve and reflexes. Invariably, the accomplice wins.

What is the accomplice's secret number, and what is the probability that someone else in the audience has it?

Interestingly, all these examples boil down to questions about the elementary structure of circulant matrices. Recall that a circulant (retrocirculant) is a square matrix identified by its top row. This row is shifted successively a step to the right (left) with wraparound to produce the remaining rows. We use the notation $\mathrm{Circ}(a, b, c, \ldots)(\mathrm{Circ}_{\leftarrow}(a, b, c, \ldots))$ where $a, b, c \ldots$ are distinct colors. Major (minor) diagonals are the monochrome sets in $\mathrm{Circ}(a, b, \ldots)(\mathrm{Circ}_{\leftarrow}(a, b, \ldots))$. The toroidal chessboard, ring dancers, and sectored disks are covered by the following basic result.

THEOREM. *It is possible to choose n positions in the $n \times n$ circulant matrix $\mathrm{Circ}(a, b, c, \ldots)$ which are* (i) *on distinct columns,* (ii) *on distinct rows,* (iii) *of distinct colors; if and only if $n \geq 1$ is odd. It is possible to replace* (iii) *with* (iii') *of distinct colors both in $\mathrm{Circ}(a, b, c, \ldots)$ and $\mathrm{Circ}_{\leftarrow}(a, b, c, \ldots)$; if and only if $n \geq 1$ is an odd integer not divisible by 3.*

Proof. Let $n \geq 1$ be odd. Superimpose $\mathrm{Circ}(0, 1, 2, \ldots, n-1)$ on $\mathrm{Circ}(a, b, \ldots)$. Let $(x)_n$ denote the nonnegative remainder after x is divided by n. The position of the kth column and a_kth row shows diagonal number $(k - a_k)_n, k = 0, 1, \ldots, n-1$. To locate the n positions which satisfy (i)–(iii), begin circling elements starting with the upper leftmost (zero). We move from left to right, as the knight does in chess, over a column and down two rows circling elements as we go. This staircase pattern wraps around at midboard and continues to the last column. The n circled positions are on distinct rows because n is odd. The corresponding diagonal sequence $\{0, n-1, n-2, \ldots, 3, 2, 1\}$ shows the circled positions have distinct color in $\mathrm{Circ}(a, b, c, \ldots)$.

To argue that n odd is necessary for (i)–(iii), suppose that in column k the a_kth row element has been chosen, $k = 0, 1, \ldots, n-1$. Since the a_k and $(k - a_k)_n$ are distinct,

$$0 \equiv_n \sum_{k=0}^{n-1} (k - a_k)_n = \sum_{k=0}^{n-1} k = \frac{1}{2}(n-1)n,$$

where $\equiv_n$ denotes congruence modulo n. This means that n must be odd.

Now consider the stronger condition (iii'). Superimpose $\text{Circ}_{\leftarrow}(0, 1, 2, \ldots, n-1)$ on $\text{Circ}_{\leftarrow}(a, b, \ldots)$. Circle the same n positions as before. In general, they will not lie on distinct minor diagonals, as exemplified by the 3×3 case. But if $n \geq 1$ is not divisible by 3, the corresponding diagonal sequence $\{(k + a_k)_n\}_{k=0}^{n-1}$ will be $\{0, 3, 6, \ldots\}$ without possibility of repetition. These same fixed positions are distinctly colored in $\text{Circ}(a, b, \ldots)$ and $\text{Circ}_{\leftarrow}(a, b, c, \ldots)$ simultaneously.

In fact, $n \not\equiv 0 \pmod 3$ is necessary for (iii'). To see this, suppose that $\{a_k\}$, $\{(k - a_k)_n\}$ and $\{(k + a_k)_n\}$, $0 \leq k \leq n-1$, are all permutations of $\{0, 1, \ldots, n-1\}$. Then,

$$\sum_{k=0}^{n-1}(k - a_k)_n^2 + \sum_{k=0}^{n-1}(k + a_k)_n^2 = 2\sum_{k=0}^{n-1}k^2$$

implies

$$0 \equiv_n \sum_{k=0}^{n-1} a_k^2 = \frac{(n-1)(2n-1)n}{6},$$

which holds if and only if n is odd and not divisible by 3.

COROLLARY 1. *It is possible to place n invulnerable queens on the cylindrical or toroidal chessboard if and only if the unwrapped $n \times n$ board satisfies $n \geq 5$, n odd and not divisible by 3.*

COROLLARY 2. *It is possible to place n invulnerable queens on an $n \times n$ chessboard for all $n \geq 4$.*

Corollary 1 follows by observing that the monochrome diagonals of $\text{Circ}(a, b, \ldots)$ $(\text{Circ}_{\leftarrow}(a, b, \ldots))$ model the major (minor) diagonals of a wrapped-around chessboard. Queens placed in the circled positions on such a board remain invulnerable when it is unwrapped. Since a queen has been placed in the upper leftmost square, pruning the first row and column leaves the remaining queens in immune formation. Thus, we have already solved the classical queens problem of Corollary 2 for odd values of $n, 3 \pm n$, and the corresponding even values $n-1$.

Completion of proof of Corollary 2. Take $n = 2m + 1 > 3$, where $3|n$. We modify the construction in [2] slightly and use circulant matrix arguments. As before, the queen placed in the upper leftmost square is deleted along with the first row and column to give the solution for $n-1$. This being done, consider the remaining $2m \times 2m$ submatrix in $\text{Circ}_{\leftarrow}(m, m+1, \ldots, n-1, 0, 1, \ldots, m-1)$, shown below.

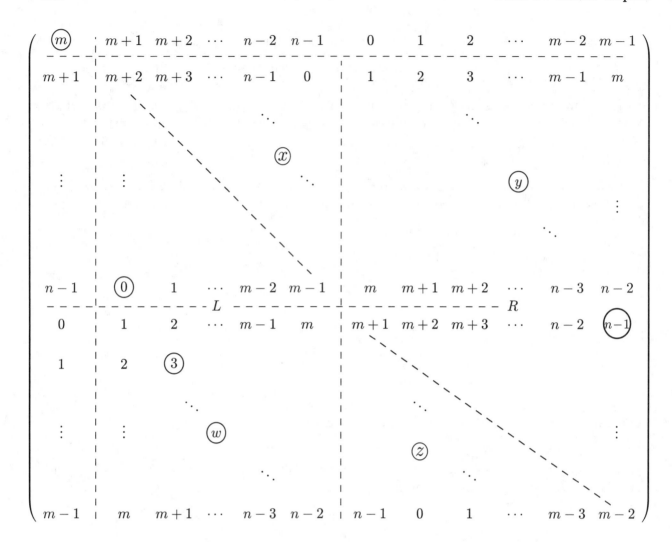

Partition this submatrix into left and right halves, denoted by L and R respectively. Place the remaining queens in positions circled as follows. Begin by circling 0 in the first column of L, then, moving from left to right, one column over and two rows down with w denoting a typical circled position in the lower half of L, $w = 3h, h = 0, 1, \ldots$. Wrap around to the top of L, where x is the typical term, $x = 3i + 1 \pmod{n}$, and continue until all m columns of L are filled. We perform the inverse process in R: Circle $n - 1$ in the last column of R, then, moving from right to left, over one column and up two rows, with y denoting a typical circled position in the upper half of R, $y = n - 3j - 1, j = 0, 1, \ldots$. Wrap around to the bottom of R, where z is the typical term, $z = n - 3k - 2 \pmod{n}$, and continue until all m columns of R are filled.

Clearly, the queens occupy distinct rows and columns. Because equations like $w = y$ have no solution $(3 \pm n - 1, 3 \pm n - 2)$, attack along minor diagonals is impossible. On the other hand, $x = z$ does have solutions, and these queens would be vulnerable on the toroidal chessboard. Here on the unwrapped board they are safe in L and R.

Confirming immunity along major diagonals is even easier. Once checks the same positions in $\text{Circ}(m - 1, m, \ldots, n - 1, 0, 1, \ldots, m - 2)$.

COROLLARY 3. *In Example (2) there is always an instant when none of the face-to-face pairs are spouses if and only if N is even. Similarly, in Example (3), there is always a superposition of disks with at least two equal sector sums if and only if n is even.*

Proof. Example (2) is absorbed into Example (3) by considering the disk problem with n *distinct* colors (= spouses) per disk. If the disks can be superimposed so that no sectors match, the ring dancers problem is solved. Let $s, t, u, \ldots$ denote either the color pattern or sequence of integer labels of the sectors of the larger disk, written linearly. Let $a, b, c, \ldots$ be the same for the smaller disk. Write $\mathrm{Circ}(a, b, c, \ldots)$, the inventory of superpositions, directly underneath $s, t, u, \ldots$. Suppose that n is even. If $s, t, u, \ldots$ are colors, move from left to right circling the element in each column of $\mathrm{Circ}(a, b, \ldots)$ which matches the color of $s, t, u, \ldots$ heading that column. If $s, t, u, \ldots$ are integers, circle $n + 1 - s, n + 1 - t, n + 1 - u, \ldots$.

Think of the circles as pigeons and the rows of $\mathrm{Circ}(a, b, \ldots)$ as the pigeonholes. In the classical (red-blue) disk problem, there would be k pigeons per column (half the column entries are color matches), giving a total of $2k^2$ pigeons filling up $2k = n$ pigeonholes. One of the rows of $\mathrm{Circ}(a, b, \ldots)$ would wind up with at least k circled elements—the desired superposition.

But here, the standard pigeonhole principle doesn't apply. Because the colors are distinct each column gets a single circle, hence, n pigeons into n pigeonholes. Enter the Theorem: The circled positions are in distinct columns and have distinct colors, but n is even. Thus, there must be a row in $\mathrm{Circ}(a, b, \ldots)$ with at least two circled elements. In the sector-sums case, this means that

$$(n + 1 - x) + x = n + 1 = (n + 1 - y) + y,$$

where x and y label two sectors of the larger disk.

For the ring dancers, the pigeonhole principle works in reverse. The two circled elements in one row imply that there must be another row *without* circled elements. That is the instant when no spouses face each other.

All these claims break down when n is odd. In that case, given colors $s, t, u, \ldots$, construct $\mathrm{Circ}(a, b, \ldots)$ by first replicating $s, t, u, \ldots$ from left to right up the minor main diagonal and then extending the colors along major diagonals. Now there is no row of $\mathrm{Circ}(a, b, \ldots)$ without a match with an element of $s, t, u, \ldots$. If $s, t, u, \ldots$ are integers and n is odd, it isn't hard to label the sectors so that every superposition gives distinct sector sums.

Example (4) is connected with circulants in a different way. Let $x_1, x_2, \ldots, x_n$ denote a circular permutation of the integers $1, 2, \ldots, n, n \geq 7$, and $s_j = x_j + x_{j+1} + x_{j+2}$. Then, $S = \mathrm{Circ}(1, 1, 1, 0, \ldots, 0)X$, where $X^T = (x_1, \ldots, x_n), S^T = (s_1, \ldots, s_n)$. Obviously, not every choice of S can result in a feasible solution X since the elements of X must be distinct integers.

LEMMA. *Nondistinct solutions of $S = \mathrm{Circ}(1, 1, 1, 0, \ldots, 0)X$ result* (i) *if two consecutive elements in S are equal; or* (ii) *if S contains five consecutive elements of the form abcba.*

Proof. The equation $x_j + x_{j+1} + x_{j+2} = x_{j+1} + x_{j+2} + x_{j+3}$ implies $x_j = x_{j+3}$, confirming (i). To confirm (ii), suppose, without loss of generality, that

$$x_1 + x_2 + x_3 = a, x_2 + x_3 + x_4 = b, x_4 + x_5 + x_6 = b, x_5 + x_6 + x_7 = a.$$

These equations imply $x_1 - x_4 = a - b = x_7 - x_4$, so $x_1 = x_7$.

The key question regarding the flim-flam game of Example (4) is: What can a player's secret number possibly be? Obviously it can't be, say, 14 because everyone has a 12 on their clockface and, next to it, at least 1 and 2. The bidding should begin at 16. The bids $17, 18, 19, 20, \ldots$ pass uneventfully.

Let σ denote any secret number. To obtain a lower bound on σ we compute $\bar{s}$, the average value of the triple-sums s_j:

$$\bar{s} = \frac{1}{12} \sum_{k=1}^{12} s_k = \frac{1}{4} \sum_{k=1}^{12} x_k = \frac{1}{4} \sum_{k=1}^{12} k = 19.5.$$

Thus $\sigma \geq 20 \geq \bar{s}$. In actuality, σ can never equal 20. Twelve integers, the largest of them 20, average out to 19.5 only if at least 6 of them equal 20. If exactly 6 of the s_j equal 20, then the others must all be 19. In that case, either two consecutive s_j are the same, or else the 20's and 19's alternate. Both possibilities are violations, in view of the above Lemma. If more than 6 of the s_j equal 20, we have the same violation because (the pigeonhole principle!) at least two consecutive entries in S must be 20.

Thus $\sigma \geq 21$. Meanwhile, my accomplice is ready with a circular permutation for which σ attains its lower bound. It is $1, 8, 10, 3, 5, 9, 4, 6, 11, 2, 7, 12$.

The only remaining question: What is the probability that someone in the audience has also stumbled onto a circular permutation of $1, 2, \ldots, 12$ for which $\sigma = 21$? Frankly, I don't know the probability, but I believe that it is very small. After all, my accomplice and I have never lost a game.

Acknowledgement. I wish to thank Professors P.J. Davis and O. Shisha for helpful discussions.

References

1. C. Berge, *Graphs and Hypergraphs*, North-Holland, 1973.
2. E.J. Hoffman, J.C. Loessi, and R.C. Moore, Constructions for the solution of the m queens problem, *Math Mag.* **42** (1969) 66–72.
3. G. Berman and K.D. Fryer, *Introduction to Combinatorics*, Academic Press, 1972.
4. D.I.A. Cohen, *Combinatorial Theory*, Wiley, 1978.
5. C.L. Liu, *Elements of Discrete Mathematics*, McGraw-Hill, 1977.
6. A. Tucker, *Applied Combinatorics*, Wiley, 1980.

Dean Clark
University of Rhode Island, Kingston
American Mathematical Monthly **92** (1985), 725–729

An Elementary Approach to the Functional Calculus for Matrices

Let A be a square matrix of complex numbers. Let $(\lambda_1 - \lambda)^{m_1} \cdots (\lambda_r - \lambda)^{m_r}$ be the minimal polynomial of A, where $\lambda_1, \ldots, \lambda_r$ is a complete enumeration of the distinct eigenvalues of A and $m_j (j = 1, \ldots, r)$ equals the largest order of the Jordan blocks with eigenvalue λ_j. Let f be a function analytic on a neighborhood of $\{\lambda_1, \ldots, \lambda_r\}$. It is known that a unique polynomial g of degree at most $m_1 + \cdots + m_r - 1$ exists that satisfies the interpolation conditions

$$g(\lambda_j) = f(\lambda_j), \ldots, g^{(m_j - 1)}(\lambda_j) = f^{(m_j - 1)}(\lambda_j), \qquad j = 1, \ldots, r.$$

The polynomial g, which depends on f, is called the Lagrange-Sylvester interpolation polynomial [1, Chapter 5] or the Hermite interpolation polynomial [3, Chapter 1] for f on the spectrum of A. A matrix $f(A)$ may then be defined by setting $f(A) = g(A)$, where the right side means the polynomial in A that is obtained by substituting A for z in $g(z)$; see [1, Chapter 5].

The aim of this note is to give an alternative and equivalent definition of $f(A)$ that is more intuitive and transparent, using the Jordan canonical form of A in a more direct way. A key to this is given by a homomorphism from an algebra of functions to a commutative algebra of upper-triangular matrices, to be stated shortly. Our purpose in presenting this material, which is known to many researchers in the subject, is to make this method more widely known among teachers of linear algebra. With our approach, we can dispense with the use of interpolation polynomials, thus making it possible to present the functional calculus for matrices to students with a modest background in matrix theory in a more direct and understandable form. Theorem 4 guarantees that the two definitions of $f(A)$ (namely, Gantmacher's and ours) agree. Our approach leads in a natural way to Dunford's integral representation of $f(A)$ (see Dunford & Schwartz [2, Chapter 7]), thus filling a gap between the elementary and advanced theories of functions of a single operator; see Theorem 5.

We will give only the definitions and the statements of the theorems that lead to a definition of $f(A)$. Except for Theorem 5, the proofs are left as exercises for the reader.

Let f be analytic on a neighborhood of $\{\lambda_1, \ldots, \lambda_r\}$. Let n denote a positive integer. By the symbol $f^*(z)$, or simply f^*, we mean the n by n upper-triangular matrix defined by:

$$f^*(z) = \begin{pmatrix} f(z) & f'(z) & f''(z)/2! & \cdots & f^{(n-1)}(z)/(n-1)! \\ & f(z) & f'(z) & & \vdots \\ & & \ddots & \ddots & f'(z) \\ 0 & & & \ddots & f(z) \end{pmatrix}.$$

This matrix appears also in the Gantmacher's treatment [1, Chapter 5], though in different contexts.

We note that for $f(z) = z$, we have

$$f^*(\lambda) = \begin{pmatrix} \lambda & 1 & & 0 \\ & \ddots & \ddots & \\ & & \ddots & 1 \\ 0 & & & \lambda \end{pmatrix} = J,$$

i.e., a single Jordan block with eigenvalue λ. If $f(z) \equiv c$ (constant), then $f^*(\lambda) = cI$, where I denotes the identity matrix.

The following key theorems, easily verifiable, state that the map $f \to f^*$ is a homomorphism from the algebra of functions analytic on a neighborhood of $\{\lambda_1, \ldots, \lambda_r\}$ to a commutative algebra of upper-triangular matrices.

Theorem 1.

(1) $(f + g)^* = f^* + g^*$,
(2) $(cf)^* = cf^*$, *where* c *is a constant,*
(3) $(fg)^* = f^*g^* = g^*f^*$,
(4) $(f/g)^* = f^*(g^*)^{-1} = (g^*)^{-1}f^*$ *if* $g(z) \neq 0$,
(5) $(1/g)^* = (g^*)^{-1}$ *if* $g(z) \neq 0$.

As an immediate corollary to Theorem 1 we have the following theorem.

Theorem 2. *Let* J *be a single Jordan block with eigenvalue* λ. *Let* f *be a rational function not having a pole at* λ, *and let* p *and* q *be co-prime polynomials such that* $f = p/q$. *Then*

$$f^*(\lambda) = p(J) \left[q(J)\right]^{-1} = \left[q(J)\right]^{-1} p(J).$$

The next theorem states that a similar results holds for an infinite power series.

Theorem 3. *Let* J *be a single Jordan block with eigenvalues* λ. *Let* $f(z) = a_0 + a_1 z + \cdots$ *be an infinite power series whose radius of convergence is strictly greater than* $|\lambda|$. *Then*

$$f^*(\lambda) = a_0 I + a_1 J + \cdots.$$

The Jordan canonical form of a square matrix A is given by

$$V^{-1} A V = \text{diag}[J_1, \ldots, J_m],$$

where V is an invertible matrix and the right side denotes the block diagonal matrix with the diagonal blocks $J_1, \ldots, J_m$. Here J_i denotes a single Jordan block with eigenvalues $\mu_i (i = 1, \ldots, m)$. The μ_i are not necessarily distinct. If f is analytic on a neighborhood of the eigenvalues $\{\mu_1, \ldots, \mu_m\}$ of A, then we can define $f(A)$ by

$$V^{-1} f(A) V = \mathrm{diag}[f^*(\mu_1), \ldots, f^*(\mu_m)].$$

Theorems 2 and 3 show that $f(A)$ is what we expect if f is either rational or a power series.

As further applications of foregoing theorems, we obtain the following two theorems. The first one gives a necessary and sufficient condition for equality of $f(A)$ and $g(A)$. The second one leads us to the Dunford's integral representation of $f(A)$.

Theorem 4 (Identity theorem). *Let* $\lambda_1, \ldots, \lambda_r$ *be distinct eigenvalues of a square matrix* A. *Let* f *and* g *be analytic on a neighborhood of* $\{\lambda_1, \ldots, \lambda_r\}$. *Then* $f(A) = g(A)$ *if and only if*

$$f^{(i)}(\lambda_j) = g^{(i)}(\lambda_j), \quad i = 0, 1, \ldots, m_j - 1, \quad j = 1, \ldots, r,$$

where m_j *denotes the largest order of the Jordan blocks with eigenvalue* λ_j.

Theorem 5 (Dunford's integral representation of $f(A)$). *Let* C *be a simple closed curve that encloses in its interior every eigenvalue of a square matrix* A. *Let* f *be analytic on* C *and in the interior of* C. *Then*

$$f(A) = \frac{1}{2\pi i} \int_C f(t)(tI - A)^{-1} dt,$$

where the right side means, by definition, the elementwise integration.

Proof. It is sufficient to prove the above equation for a single Jordan block J of order k with eigenvalue λ_j. We have

$$f(J) = f^*(\lambda_j) \qquad \text{(by Theorems 2 and 3)}$$

$$= \frac{1}{2\pi i} \int_C f(t) \begin{pmatrix} \dfrac{1}{t - \lambda_j} & \dfrac{1}{(t - \lambda_j)^2} & \cdots & \dfrac{1}{(t - \lambda_j)^k} \\ & \ddots & & \vdots \\ & & \ddots & \dfrac{1}{(t - \lambda_j)^2} \\ 0 & & \ddots & \dfrac{1}{t - \lambda_j} \end{pmatrix} dt$$

$$= \frac{1}{2\pi i} \int_C f(t)(tI - J)^{-1} dt.$$

References

1. F.R. Gantmacher, *The Theory of Matrices*, Vol. 1, Chelsea Publishing Company, New York, 1959.

2. N. Dunford and J.T. Schwartz, *Linear Operators Part I: General Theory*, Wiley Interscience, New York, 1958.
3. B. Wendroff, *Theoretical Numerical Analysis*, Academic Press, New York, 1966.

Yasuhiko Ikebe and Toshiyuki Inagaki
University of Tsukuba
American Mathematical Monthly **93** (1986), 390–393

PART 10
Problems

Introduction

To best understand linear algebra, one must work problems—the more, the better. Hence, any material directed toward the learning of linear algebra should include a selection of problems. This section contains a collection we have accumulated from several submitters, ourselves, and the problem sections of the *American Mathematical Monthly*. They are roughly divided into sections corresponding to those of this volume. The problems vary greatly in difficulty and subtlety, and we have not attempted to grade this difficulty. However, often a simple idea will make a seemingly difficult problem transparent. Many problems are also good exercises for the elementary course.

Most problems are attributed immediately following the statement. If the problem is from the *Monthly* and we know of the appearance of a solution, the solution (only) is referenced as "soln [year, page number(s)]." Other problems from the *Monthly* are referenced by "prob [year, page number(s)]." Submitted problems are attributed to the submitter, where possible, and the few problems not referenced are either unattributable or from the Editors.

Problems

Partioned Matrix Multiplication

1. Let A be an $m \times r$ matrix, B $r \times n$ and C $n \times s$. Determine a formula that gives the number of scalar multiplications in computing AB. Use this result to derive a formula for the number of scalar multiplications in computing $(AB)C$ and $A(BC)$.

   Find the most and least efficient ways of performing the following products. A is 5×14, B is 14×87, C is 87×3, D is 3×42.

   (a) ABC

   (b) *ABCD*.

   Gareth Williams
   Stetson University

2. Let A, m_1-by-n_1, and C, m_2-by-n_1, be given real matrices (any field will do) and let X, m_2-by-n_2, be free to be chosen. Discover and prove a formula for the maximum value of

$$\text{rank} \begin{bmatrix} A & B \\ C & X \end{bmatrix},$$

   in which the minimum is taken over all choices of X. Your formula should be in terms of

$$\text{rank}\,[A, B]\,, \quad \text{rank} \begin{bmatrix} A \\ C \end{bmatrix}, \quad \text{and rank}\,[A]\,.$$

   Besides stressing the concept of rank, which is very important for work beyond the first course, this problem, though challenging, can be done in several ways by entirely elementary means. In addition, it provides entry to the unsolved problem in which the unspecified entries are scattered arbitrarily, rather than lying in the block X.

   The Editors
   CMJ 23 (1992), pp. 299–303

3. Let the rows of B be $\mathbf{r}_1, \mathbf{r}_2, \mathbf{r}_3$. You want to premultiply B to obtain

$$\begin{bmatrix} \mathbf{r}_1 + \mathbf{r}_2 \\ \mathbf{r}_2 - 2\mathbf{r}_3 \\ \mathbf{r}_1 + \mathbf{r}_3 - 2\mathbf{r}_2 \end{bmatrix}.$$

What is the premultiplier matrix?

Steve Maurer
Swarthmore College

4. Consider the product $AB\mathbf{x}$, where A and B are square matrices and $\mathbf{x}$ is a column vector. Compare the number of computations to compute the product in the forms $(AB)\mathbf{x}$ and $A(B\mathbf{x})$.

Steve Maurer
Swarthmore College

Determinants

5. Let an $n \times n$ matrix have positive entries along the main diagonal, and negative entries elsewhere. Assume that it is normalized so that the sum of each column is 1. Prove that its determinant is greater than 1.

Melvin Hausner
New York University
soln [1964, 447–483]

6. Given an $n \times n$ matrix with randomly selected integer elements, what is the probability that the absolute value of the determinant of the matrix is an odd integer?

Harry Lass
California Institute of Technology
soln [1965, 191]

7. If $(b) = [b_{ij}]$ is a real symmetric square matrix with $b_{ii} = 1$, and $\sum_{j \neq i} |b_{ij}| \leq 1$ for each i, then $\det(b) \leq 1$.

L.A. Shepp
Bell Telephone Laboratories
soln [1966, 1024–1025]

8. Find necessary and sufficient conditions for a $k \times n$ matrix ($k < n$) with integral elements in order that it be a submatrix of an integral $n \times n$ matrix with determinant 1.

Dorembus Leonard
Tel-Aviv University
soln [1969, 300]

9. Let A be a complex $n \times n$ matrix, let $\overline{A}$ be its complex conjugate, and let I be the $n \times n$ identity matrix. Prove that $\det(I + A\overline{A})$ is real and nonnegative.

D.Ž. Djoković
University of Waterloo
soln [1976, 483–484]

10. Let A, B, C, D be $n \times n$ matrices such that $CD' = DC'$, where the prime denotes transpose. Prove

$$\begin{vmatrix} A & B \\ C & D \end{vmatrix} = |AD' - BC'|.$$

Anon, Erewhon-upon-Yarkon
soln [1977, 495–496]

11. Suppose A, B, C, D are n by n matrices and C', D' denote the transposes of C, D. In a comment on problem 6057 [1987, 1020], M.J. Pelling remarked that the condition

$$CD' + DC' = 0 \qquad\qquad (*)$$

implies

$$\det \begin{pmatrix} A & B \\ C & D \end{pmatrix} = \det(AD' + BC') \qquad\qquad (**)$$

if D is nonsingular, but gave an example with singular D satisfying $(*)$ in which one side of $(**)$ is 1 and the other side is -1. Prove that in any case $(*)$ implies

$$\det \begin{pmatrix} A & B \\ C & D \end{pmatrix}^2 = \det(AD' + BC')^2.$$

William C. Waterhouse
Pennsylvania State University, State College
soln [1990, 244–250]

12. Let A_n be the matrix of order $(2^n - 1) \times n$ whose kth row is the binary expression for k. Let $M_n = A_n A_n' (\mathrm{mod}\, 2)$. If M_n is regarded as a matrix over the integers, what is its determinant?

Stephen B. Maurer
Princeton University
soln [1977, 573–574]

13. Let A be the cyclic matrix with $(a_0, a_1, \ldots, a_{p-1})$ as first row, p a prime. If a_i are integers, show that $\det A \equiv a_0 + a_1 + \cdots + a_{p-1} (\mathrm{mod}\, p)$.

Ira Gessel
Massachusetts Institute of Technology
soln [1979, 129–130]

14. If m and n are given positive integers and if A and B are m by n matrices with real entries, prove that $(\det AB^T)^2 \le (\det AA^T)(\det BB^T)$.

S.J. Bernau and Gavin G. Gregory
University of Texas at El Paso
soln [1994, 81]

15. In elementary linear algebra, two different definitions of the word "adjoint" are used. The adjoint of a square matrix A with complex entries is either:

    (I) the matrix whose (i, j)-entry is the cofactor of a_{ij} in A; or,

    (II) the complex conjugate of the transpose of A.

    Under what conditions on the matrix A will these two definitions yield the same matrix?

    Richard Sinkhorn
    University of Houston
    soln [1993, 881–882]

16. Evaluate the determinant

$$
\begin{vmatrix}
x & x-1 & \cdots & x-k \\
x-1 & x & \cdots & x-k+1 \\
\cdot & \cdot & \cdots & \cdot \\
x-k & x-k+1 & \cdots & x
\end{vmatrix}.
$$

    H.S. Shapiro
    Chatham, N.J.
    soln [1953, 553]

17. Let A and B be two nth-order determinants, and let C be a third determinant whose (i, j)th element is the determinant A with its ith column replaced by the jth column of B. Show that $C = A^{n-1}B$.

    James Ax and Lawrence Shepp
    Polytechnic Institute of Brooklyn
    soln [1958, 288]

Eigenanalysis

18. Let $A = (a_{ij})$ be an m by n matrix whose entries are elements of an ordered set $(S, \geq)[sic]$. Suppose A is column ordered—that is $a_{1j} \geq a_{2j} \geq \cdots \geq a_{mj}$ for each $j = 1, 2, \cdots, n$. Obtain a row ordered matrix A' by arranging the entries of each row of A so that $a'_{i1} \geq a'_{12} \cdots \geq a'_{in}$ for each $i = 1, 2, \cdots, m$. Is A' column ordered?

    R.W. Cottle
    University of California
    soln [1963, 212–213]

19. A square matrix M of order n has the properties that $a_{ii} = 1$ and $a_{ik}a_{kj} = a_{ij}$ for all i, j, k. What are the characteristic values of M?

    F.D. Parker
    State University of New York
    soln [1965, 321–322]

20. Show that if A and B are nonnegative definite Hermitian matrices of order n, the characteristic roots of AB are real and nonnegative.

    J.T. Fleck and Carl Evans
    Cornell Aeronautics Laboratory
    soln [1966, 791–792]

21. Suppose $a_{11}, \cdots, a_{1n}$ are given integers whose greatest common divisor is 1, and suppose $n \geq 2$. Is it always possible to find a matrix (a_{ij}) with the given integers in the first row and all a_{ij} integers such that $\det(a_{ij}) = 1$?

    B.R. Toskey
    Seattle University
    soln [1968, 81]

22. Let A_{nn} be a real, symmetric, positive semi-definite matrix. Let A_{pn} be the matrix obtained from A_{nn} by omitting the last $n - p$ rows, and A_{pp} be obtained by omitting the last $n - p$ columns of A_{pn}. Prove that rank $A_{pp} =$ rank A_{pn}.

    F.D. Faulkner
    U.S. Naval Postgraduate School
    soln [1967, 877–878]

23. If A is a normal matrix (i.e., A commutes with its conjugate transpose), then the characteristic roots of A form a symmetric set with respect to the origin in the complex plane (i.e. if Z is a characteristic root of multiplicity r then $-Z$ is a characteristic root of multiplicity r) if and only if the trace $(A^{2k+1}) = 0$ for $k = 0, 1, 2, \cdots$.

    C.M. Petty and W.E. Johnson
    Lockheed Aerospace Sciences Laboratory
    soln [1968, 205–206]

24. It is well known that each characteristic vector for a nonsingular $n \times n$ matrix A is also a characteristic vector for the matrix adj A of (transposed) cofactors of elements of A. Prove that this is true for singular A also.

    D.E. Crabtree
    Amherst College
    soln [1968, 1127–1128]

25. Suppose all three matrices, $A, B, A + B$ have rank 1. Prove that either all the rows of A and B are multiples of one and the same row vector v or else all the columns of A and B are multiples of one and the same column vector w.

    Peter Ungar
    New York University
    soln [1982, 133]

26. An $n \times n$ complex matrix $A = [a_{ij}]$ is cross-diagonal if $a_{ij} = 0$ whenever $i + j \neq n + 1$. Find the condition that the eigenvectors of A span $\mathbb{C}^n$, the entire n-vector space.

H. Kestelman
University College, London
soln [1986, 566]

27. A and B are $n \times n$ matrices and both have only positive eigenvalues; can AB have only negative eigenvalues? How is the result affected if A, B are both Hermitian?

H. Kestelman
University of College, London
soln [1984, 587–588]

28. (a) For what positive integers n does there exist an n by n matrix A over $\mathbb{C}$ having the following three properties:

   (i) $n^2 - n$ of the entries are zero,

   (ii) there are n distinct nonzero entries $r_1, r_2, \ldots, r_n$, none of which lies on the main diagonal,

   (iii) the eigenvalues of A are $r_1, r_2, \ldots, r_n$?

   (b) Same as (a) except that all the entries are required to be real.

Bruce A. Reznick and Lee A. Rubel
University of Illinois, Urbana
soln [1989, 532]

29. Let A be a matrix (not necessarily square) of rank $r \geq 1$ with nonnegative elements. Let A^* denote the transpose of A. Prove that the square matrix AA^* has no eigenvalue different from 0,1 if and only if, after deleting all identically vanishing rows and columns, the remaining submatrix of A can be brought by a permutation of the rows and a permutation of the columns to the form

$$(1) \qquad\qquad B = B_1 + \cdots + B_r,$$

where, for each i, B_i is a rectangular matrix of rank 1, with all its elements positive and such that the sum of squares of all elements of B_i is 1. Equation (1) means that B is obtained by laying out successively the rectangular blocks $B_1, \cdots, B_r$ with the lower right corner of B_i attached to the upper left corner of B_{i+1} and with zeros filling in the entire matrix outside these blocks.

Ky Fan
Wayne State Uniersity
soln [1961, 1011–1012]

30. For a timed paper and pencil test on eigentheory, it is only fair that the eigenvalues and eigenvectors should be simple but unguessable. What's a good way for a professor to create matrices A with this property? (Trial and error is not a good way—try it.)

Use your method to create a simple 2×2 matrix with small-integer but unguessable eigenvalues and eigenvectors.

Steve Maurer
Swarthmore College

Geometry

31. The norm $\|A\|$ of a real 2×2 matrix A is by definition the maximum of $\|A\hat{x}\|$ when $\hat{x}\| = 1$; if $\|x\|$ is the euclidean norm $(x^Tx)^{1/2}$, then $\|A\| < \||A|\|$ where $|A|$ is the matrix whose elements are the absolute magnitudes of those of A. Find necessary and sufficient conditions on an invertible 2×2 matrix N in order that $\|A\| < \||A|\|$ for all A when $\|x\|$ is defined as the euclidean norm of N_x.

    (One tends to use the inequality $\|A\| < \||A|\|$ automatically in matrix analysis and might "naturally" assume it, when $\|x\|$ is the euclidean norm of N_x, for all N.)

    H. Kestelman
    University College, London
    soln [1981, 831]

32. Let A be a nonnegative irreducible $n \times n$ matrix with Perron root r. Show that there exists a constant $k > 0$ such that

$$\left\|(tI - A)^{-1}\right\| \le \frac{K}{t - r} \text{ for all } t > r.$$

    Here I denotes the identity matrix and $\| \ \|$ is the operator norm induced by the Euclidean norm $x \to (x^Tx)^{1/2}$. Find the best value of the constant K.

    Emeric Deutsch
    Polytechnic Institute of New York
    soln [1983, 214–215]

33. Given that the sequence $A, A^2, A^3, \ldots$ converges to a nonzero matrix A^∞, show that $A^\infty = V(WV)^{-1}W$ where V is any matrix whose columns are a basis of the right kernel of $A - I$, and W is any one whose rows are a basis of the left kernel of $A - I$.

    H. Kestelman
    University College, London
    soln[1983, 485]

34. Given a positive integer n, let M denote the set of n by n matrices with complex entries. If $A \in M$, let A^* denote the conjugate transpose of A. Prove that for each positive integer k and each $B \in M$ there exists a unique $A \in M$ such that $A(A^*A)^k = B$.

    Bjorn Poonen
    Winchester, MA
    soln [1991, 763]

35. Let k and n be integers with $0 < k < n$, and let A be a real n by n orthogonal matrix with determinant 1. Let B be the upper left k by k submatrix of A, and let C be the lower right $(n-k)$ by $(n-k)$ submatrix of A.

    (a) Show that $\det(B) = \det(C)$.

    (b) Give a geometrical interpretation.

    (c) Generalize to the case in which A is a unitary matrix.

    M. Al-Ahmar
    Al-Fateh University, Tripoli, Libya
    prob 10279 [1993, 76]

36. A square $n \times n$ matrix $A = (a_{ik})$ with complex numbers as elements is called normal if $AA^* = A^*A$, where A^* is the transposed and complex conjugate of A. A matrix A is called nilpotent if for some integer $r \geq 1$ the matrix A^r is the zero matrix 0. Prove (by rational methods only) that a normal and nilpotent matrix is the zero matrix.

    Olga Taussky
    National Bureau of Standards
    soln [1958, 49]

37. A square matrix B is called normal if its elements are in the complex field and if it commutes with its transposed conjugate matrix B^* (i.e., if $BB^* = B^*B$); more generally, call a given $n \times n$ matrix A m-normal if A can be imbedded, as leading $n \times n$ submatrix, in some normal $m \times m$ matrix B. Find the greatest value of n for which every (complex) $n \times n$ matrix is $(n+1)$-normal, and show that every square matrix is m-normal for all sufficiently large m.

    What are the corresponding results when A, B are restricted to be real?

    M.P. Drazin
    RIAS, Baltimore, MD
    soln [1060, 383–385]

Matrix Forms

38. F is a field of characteristic $\gamma(\gamma \neq 0)$; T is a $p \times p$ nonsingular matrix over F. Prove that the only matrix of the form $\lambda T(\lambda \in F)$ similar to T is T itself.

    Ih-Ching Hus
    Fordham University
    soln [1969, 418–419]

39. Two matrices A and B are *permutation equivalent* if B can be obtained from A by first permuting the rows of A and then permuting the columns of the resulting matrix.

Call an $n \times n$ matrix of zeros and ones a $k - k$ matrix if there are precisely k ones in each row and in each column. Show that if $n \leq 5$, then every $k - k$ matrix is permutation equivalent to its transpose, but that this is no longer true if $n \geq 6$.

Morris Newman and Charles Johnson
National Bureau of Standards
soln [1976, 201–204]

40. Given a complex square matrix A, show that there exists a unitary matrix U such that U has all diagonal entries equal. If A is real, U can be taken real orthogonal.

H.S. Witsenhausen
Bell Laboratories, Murray Hill, NJ
soln [1980, 62–63]

41. Let M be a 3 by 3 matrix with entries in a field F. Prove that M is similar over the field to precisely one of these three types:

$$\begin{pmatrix} a & 0 & 0 \\ 0 & a & 0 \\ 0 & 0 & a \end{pmatrix} \quad \begin{pmatrix} b & 0 & 0 \\ 1 & c & 0 \\ 0 & 0 & c \end{pmatrix} \quad \begin{pmatrix} d & 1 & 0 \\ e & 0 & 1 \\ f & 0 & 0 \end{pmatrix} \quad (a, b, c, d, e, f \in F).$$

$$\text{Type I} \qquad \text{Type II} \qquad \text{Type III}$$

F.S. Cater
Portland State University
soln [1987, 881–882]

42. Let A be a partitioned real matrix with submatrices $A_{ij}, i, j = 1, 2, \ldots, k$. Let B be the $k \times k$ matrix with elements b_{ij} given as follows: b_{ij} is the algebraic sum of all of the elements in the A_{ij} submatrix. Show that if A is symmetric positive definite, then B is symmetric positive definite.

William N. Anderson, Jr., and George E. Trapp
Fairleigh Dickinson University and West Virginia University
soln [1988, 261–262]

43. Prove that the product of two skew-symmetric matrices of order $2N$ has no simple eigenvalues.

M. Lutzky
Silver Spring, MD

44. In a 6×7 reduced echelon matrix with five pivotal 1's, at most how many entries are nonzero? How about in an $m \times n$ echelon matrix with p pivots?

45. In a 6×7 echelon matrix with five pivotal entries, at most how many entries are nonzero? How about an $m \times n$ echelon matrix with p piviots?

46. Give an example of a 3×3 lower triangular *echelon* matrix.

Matrix Equations

47. If A, B are invertible matrices of the same dimension, it is not always possible to solve $XY = A, YX = B$ for X, Y; A and B must be similar, since $X^{-1}AX = B$. Under what conditions on invertible A, B, C can one solve $XY = A, YZ = B, ZX = C$ for X, Y, Z?

J.L. Brenner
Stanford Research Institute
soln [1962, 166–167]

48. Let m and n be positive integers. What pairs of matrices C and D, over any field K, have the property that if A is an $m \times n$ matrix over K and B is an $n \times m$ matrix over K such that $AB = C$ then $BA = D$?

William P. Wardlaw
U.S. Naval Academy
soln [1981, 154]

49. Suppose A, B, and C are matrices of size m by n, m by m, and n by m respectively and suppose CA is equal to the n by n identity matrix. Give a necessary and sufficient condition for the block matrix $M = \begin{pmatrix} A & B \\ 0 & C \end{pmatrix}$ to be invertible and find an expression for M^{-1} when this condition holds.

Lawrence A. Harris
University of Kentucky
soln [1992, 60]

50. Let C be an $n \times n$ matrix such that whenever $C = AB$ then $C = BA$. What is C?

Harley Flanders
University of California
soln [1954, 470]

51. Let A, B, X denote $n \times n$ matrices. Show that a sufficient condition for the existence of at least one solution X of the matrix equation

$$X^2 - 2AX + B = 0$$

is that the eigenvalues of the $2n \times 2n$ matrix

$$R = \begin{bmatrix} A & I \\ A^2 - B & A \end{bmatrix}$$

be pairwise distinct. Here I denotes the $n \times n$ identity matrix.

Peter Treuenfels
Brookhaven National Laboratory
soln [1059, 145–146]

Linear Systems, Inverses and Rank

52. Let A and B be arbitrary matrices of dimensions $m \times n$ and $n \times m$ respectively. Does the existence of $[I_m + AB]^{-1}$ imply the existence of $[I_n + BA]^{-1}$, where I_m and I_n are the identity matrices of dimension m and n respectively? If so, determine the latter in terms of the former. If not, give a counterexample.

    Henry Cox and David Taylor
    Model Basin, Washington, DC
    soln [1966, 543–544]

53. Let a_{ij} be nonnegative real numbers such that (a) $a_{ii} = 1$, (b) $a_{ij}a_{ji} < 1$, and (c) $a_{ij}a_{jk} \le a_{ik}$ for all i, j with $1 \le i, j \le n$, and $i \ne j$.

    Show that the matrix $[a_{ij}]$ has a positive determinant for $n = 3$. This is obviously true for $n = 1, 2$; can you establish this result for $n > 3$? At least show that $[a_{ij}]$ is non-singular for all n.

    J.C. Nichols and (independently) J.A. Huckaba
    Monmouth College and University of Missouri
    soln [1970, 530–531]

54. Show that the square matrix $M = (m_{ij})$ is nonsingular if it satisfies the following conditions:

    (i) $m_{ii} \ne 0$ for all i;

    (ii) if $i \ne j$ and $m_{ij} \ne 0$, then $m_{ji} = 0$;

    (iii) if $m_{ij} \ne 0$ and $m_{jk} \ne 0$, then $m_{ik} \ne 0$.

    R.D. Whittekin
    Metropolitan State College
    soln [1975, 938–939]

55. If Q_1, Q_2, Q_3, Q_4 are square matrices of order n with elements in a field F and the matrix $\begin{bmatrix} Q_1 & Q_3 \\ Q_2 & Q_4 \end{bmatrix}$ of order $2n$ is invertible, are there scalars $\alpha_1, \alpha_2, \alpha_3, \alpha_4$ in F so that $U = \alpha_1 Q_1 + \alpha_2 Q_2 + \alpha_3 Q_3 + \alpha_4 Q_4$ is invertible?

    Robert Hartwig
    North Carolina State University
    soln [1983, 120]

56. Suppose A and B are n by n matrices and suppose there exists a nonzero vector x such that $Ax = 0$ and a vector y such that $Ay = Bx$. If A_i is the matrix obtained from A by replacing its ith column by the ith column of B, show that

$$\sum_{i=1}^{n} \det A_i = 0.$$

    Tim Sauer
    George Mason University
    soln [1990, 245–246]

Problems 57–72 submitted by Steve Maurer, Swarthmore College

57. Suppose Gaussian elimination reduces A to

$$R = \begin{bmatrix} 1 & 2 & 0 & -3 \\ 0 & 0 & 1 & 1 \\ 0 & 0 & 0 & 0 \end{bmatrix}.$$

a) Name all solutions to $A\mathbf{x} = 0$.

b) But you were never told the entries of A! What theorem justifies your answer to part a?

c) Suppose further that Gaussian elimination required no row switches to reach R. Find a vector $\mathbf{b}$ such that $A\mathbf{x} = \mathbf{b}$ has no solutions. *Caution*: If all you show about your $\mathbf{b}$ is that $R\mathbf{x} = \mathbf{b}$ has no solutions, you are done.

58. Every real number except 0 has a multiplicative inverse. Does every square matrix that is not all 0's have an inverse? Does every square matrix that has no 0 entries have an inverse?

59. Let A be $n \times n$. How much work does it take to solve $A\mathbf{x} = \mathbf{b}$ by first finding A^{-1} and then premultiplying $\mathbf{b}$ by it? How much more work is this than doing Gaussian elimination? (Computing A^{-1} by Gaussian elimination takes approximately n^3 steps.)

60. Your boss gives you a huge square matrix A and tells you to solve $A\mathbf{x} = \mathbf{b}$. Just as you are about to start, your fairy godmother arrives and hands you A^{-1} on a silver platter. To solve the problem most efficiently, what should you do?

61. Repeat the previous problem, but this time it so happens that yesterday you solved $A\mathbf{x} = \mathbf{c}$ by the LU method and you've saved the L and U matrix. Now what should you do?

62. It is often stated as a theorem that

$$(AB)^{-1} = B^{-1}A^{-1}.$$

This equality is misleading because it suggests a bi-implication—that the existence of both inverses on the right implies the existence of the inverse on the left *and* vice versa. However, vice versa is false: A and B don't even have to be square (hence certainly not invertible) and yet AB can be square and invertible.

a) Construct a specific example confirming the previous sentence.

b) Prove: If AB is invertible and A is square, then both A and B are invertible and $(AB)^{-1} = B^{-1}A^{-1}$.

63. The usual definition of A being invertible is that there exists a matrix A^{-1} such that $AA^{-1} = A^{-1}A = I$. From this it follows trivially that a nonsquare matrix A can't have an inverse, because $A^{-1}A$ and AA^{-1} wouldn't even have the same shape, let alone be equal. What if we change the definition of inverse and ask only that $A^{-1}A$ and AA^{-1} both be identity matrices, maybe with different sizes? Prove or disprove: Now some nonsquare matrices have inverses.

64. Prove: If A is square and has a right inverse, then it has an inverse. (*Hint:* What conclusion does the hypothesis force concerning the reduced echelon form of A?)

65. Prove or disprove: Let $\mathbf{u}, \mathbf{v}, \mathbf{w}$ be vectors in some space V. If the sets $\{\mathbf{u}, \mathbf{v}\}, \{\mathbf{u}, \mathbf{w}\}$, and $\{\mathbf{v}, \mathbf{w}\}$ are each independent, then the single set $\{\mathbf{u}, \mathbf{v}, \mathbf{w}\}$ is independent.

66. If a matrix A has a unique right inverse, then A' is invertible.

67. Suppose A and $B = [B_1|B_2]$ are both matrices with m rows and independent columns. Prove that this matix has independent columns.

$$\begin{bmatrix} A & B_1 & \mathbf{0} \\ A & \mathbf{0} & B_2 \end{bmatrix}.$$

68. Consider the matrix equation

$$A\mathbf{x} = \mathbf{b} \tag{1}$$

and the single linear (dot product) equation

$$\mathbf{c} \cdot \mathbf{x} = d. \tag{2}$$

Suppose (1) $\implies$ (2) in the logical sense that any vector $\mathbf{x}$ that satisfies (1) satisfies (2). Prove that the implication can be proved algebraically by simply summing the right linear combination of the rows of (1). In other words, there is a row vector $\mathbf{v}$ such that

$$\mathbf{v}A = c \qquad \text{and} \qquad \mathbf{v}\mathbf{b} = d.$$

Put another way, for linear equations "logically implies" is no stronger than "implies by linear calculations."

69. An *inconsistent* system $A\mathbf{x} = \mathbf{b}$ has 13 equations and 17 unknowns. What is the largest possible rank of A? Explain briefly.

70. Three sets of variables are related as follows:

$$y_1 = -x_1 + 2x_2, \qquad z_1 = 2y_1 + 3y_2,$$
$$y_2 = x_1 - x_2, \qquad z_2 = y_1 - 4y_2.$$

Use matrix operations to obtain formulas for the z's in terms of the x's.

71. Correct the wording in the following statements.

a) A basis of a matrix is the columns with pivots.

b) (1,0) and (0,1) are the span of R^2.

c) $\begin{bmatrix} 1 & 2 & 1 \\ 3 & 6 & 2 \end{bmatrix} = \begin{bmatrix} 1 & 2 & 1 \\ 0 & 0 & -1 \end{bmatrix}$, so the matrix has no solutions.

72. The matrix equation $A\mathbf{x} = \mathbf{0}$ represents the intersection of four planes in R^3 and this intersection is a line. What is the rank of matrix A and what is its nullity?

73. Instructions: Please work with a partner and turn in a solution paper. Part (a) is easy but you may need to think some how to answer part (b). Discuss this with your partner until you figure out a method, and then use MATLAB to do any necessary calculations.

   Briefly but clearly explain your method for each part and why it works. No credit will be given unless the method is valid and justified correctly. Be sure you both agree that the explanations you submit are clear and correct. Your solution paper should also include the results of any calculations you do.

   Use the following matrices:

$$A = \begin{bmatrix} 18 & -13 & -4 & -2 \\ 12 & -21 & -11 & 8 \\ -34 & 5 & 3 & 34 \\ 22 & -9 & -2 & -32 \\ -24 & -8 & 2 & 4 \end{bmatrix} \qquad B = \begin{bmatrix} -1 & -1 & -3 & 3 & -3 \\ 0 & -6 & 4 & 5 & 0 \\ 5 & 0 & 6 & -3 & 1 \\ -5 & 1 & -5 & 3 & -1 \\ 0 & 2 & 2 & 0 & 0 \end{bmatrix}$$

   a) Show that Col A and Col B have the same dimension.

   b) Determine whether or not Col A and Col B are the same subspace of R^5.

   (*Remarks*: This is a nontrivial question. For example, if two subspaces of R^5 each have dimension 1, we could visualize each as a line through the origin, but they might not be the same line. If each has dimension 2, they look like planes through the origin, but they might not be the same plane. If each has dimension 3, each looks like R^3—but they might not be the same sets, etc. Your job here is to figure out a way to decide if two subspaces of R^5, of the same dimension, are actually the same set of vectors, and apply that method to the subspaces Col A and Col B.)

   Jane Day
   San Jose State University

74. Compute the reduced echelon forms of a number of 3×3 matrices entered at random. Why do you get these reduced echelon forms? How can you get another type of reduced echelon form?

   Gareth Williams
   Stetson University

Applications

75. An $n \times n$ array (matrix) of nonnegative integers has the property that for any zero entry, the sum of the row plus the sum of the column containing that entry is at least n. Show that the sum of all elements of the array is at least $n^2/2$.

   S.W. Golomb and A.W. Hales
   California Institute of Technology and King's College, Cambridge, England
   soln [1963,1006]

76. Let $A = (a_{ij})$ be a real $n \times n$ matrix with $a_{ij} \geq 0, 1 \leq i, j \leq n$. Prove that $r(A) \leq r\left[\frac{1}{2}(A + A^T)\right]$, where $r(C)$ denotes the spectral radius of a matrix C.

    B.W. Levinger
    Case Western Reserve University
    soln [1970, 777–778]

77. Prove or disprove: If $A = (a_{ij})$ is a nonsingular $n \times n$ complex matrix, if also $A^{-1} = (b_{ij})$, and each a_{ij} and b_{ij} is nonzero, then the matrices of reciprocals (a_{ij}^{-1}) and (b_{ij}^{-1}) are singular or nonsingular together. H. Flanders [*American Mathematical Monthly* (1966), 270–272] proved this for $n = 3$.

    P.M. Gibson
    University of Alabama at Huntsville
    soln [1971, 405–406]

78. Find all positive semi-definite Hermitian matrices $A = (a_{ij})$ with the property that the matrix of reciprocals $(1/a_{ij})$ is also positive semi-definite.

    Gérard Letac
    Université de Clermont, France
    soln [1975, 80–81]

79. Let $A = (a_{ij})$ be a real square matrix such that $a_{ij} > 0$ for $i \neq j$. Show that all entries of e^A are positive.

    Melvin Hausner
    Courant Institute, New York University

    Problems 80–83 submitted by Steve Maurer, Swarthmore

80. The LU factorization of a certain matrix A is

$$\begin{bmatrix} 2 & 0 & 1 \\ 1 & 1 & -1 \\ 2 & 3 & 3 \end{bmatrix}.$$

    a) What is the solution to

$$A\mathbf{x} = \begin{pmatrix} 4 \\ 1 \\ 5 \end{pmatrix}?$$

    b) What is A?

81. Two planes fly along straight lines. At time t plane 1 is at $(75, 50, 25) + t(5, 10, 1)$ and plane 2 is at $(60, 80, 34) + t(10, 5, -1)$.

    a) Do the planes collide?

    b) Do their flight paths intersect?

82. Suppose that the center of a sphere S having radius 3 units is moving at constant velocity through space along a line that passes through the point $P_0 = (2, 2, 2)$, and is parallel to the vector $(1, -2, 2)$. If the center of the sphere is at P_0 at a certain time, and the sphere collides with the plane T, having the equation $x - 2y + 2z = 15$, two seconds later, then:

a) What is the direction of motion of S?

b) What point of T is hit by S?

c) With what velocity was S moving prior to the collision?

d) What is the equation of motion of the ball after the collision? (Assume that the collision occurs at time $t = 0$ seconds, the angle of incidence equals the angle of reflection, the sphere's center moves in the same plane before and after the collision, and the velocity is the same both before and after. Give a formula for the position of the sphere's center t seconds after the collision.)

(Parts a–c of this problem appeared on the final exam at the University of Waterloo in 1973; part d is an addition.)

83. (Simpson's Rule in calculus) This problem shows how it is easy, using linear algebra concepts, to see that there must be an integration method like Simpson's and also easy to find the specific constants. (Similar reasoning leads to other Simpson-like methods as well.)

Let $\mathcal{P}_n$ be the vector space of polynomials of degree at most n. Let a_0, a_1, a_2 be three distinct real numbers.

a) Show that for any triple (y_0, y_2, y_2) there is a unique $p \in \mathcal{P}_2$ such that $p(a_0) = y_0, p(a_1) = y_1$ and $p(a_2) = y_2$. Show that the function T that maps each triple to its polynomial is linear. [Hint: $T = E^{-1}$, where $E : \mathcal{P}_2 \to R^3$ is the evaluation map $E(p) = (p(a_0), p(a_1), p(a_2))$.] Show that E is linear and invertible.

b) Show that the integration map $S(f) = \int_{a_0}^{a_2} f(x)dx$ is a linear transformation from any space of continuous functions (e.g., $\mathcal{P}_2$).

c) Why must the composition $SE^1 : R^3 \to R$ have a formula of the form $SE^{-1}(y_0, y_1, y_2) = c_0 y_0 + c_1 y_1 + c_2 y_2$?

d) We want to evaluate the integral $S(f)$. If our evaluation is exact for $p \in \mathcal{P}_2$, it is likely to be close for most continuous functions, so long as the interval $[a_0, a_2]$ is small. Use the fact that $S = (SE^{-1})E$ to show that for $p \in \mathcal{P}_2$

$$\int_{a_0}^{a_2} p(x)dx = c_0 p(a_0) + c_1 p(a_1) + c_2 p(a_2) \tag{1}$$

for the same constants c_0, c_1, c_2 as in part c. In other words, there has to be a Simpson-like rule for evaluating integrals.

e) Find c_0, c_1, c_2 if $a_1 = a_0 + h$ and $a_2 = a_1 + h$. It turns out the c's depend only on h, not on the a's. Since equation (1) holds for all $p \in \mathcal{P}_2$, it suffices to plug in p's for which both sides are simple. Suggestion: Use $p_1(x) = 1, p_2(x) = x - a_1$, and $p_3(x) = [(x - a_1)/h]^2$.

f) Derive Simpson's Rule; you must merely string together a number of intervals of the form in part (e).

g) For any vector space V of continuous functions on interval $[a, b]$, show that $\int_a^b f(x)dx = \sum c_i p(a_i)$ is true for all $f \in V$ if and only if it is true for a basis of V. Consequently, prove that equation (1) is true for all $p \in \mathcal{P}_3$ (not just $\mathcal{P}_2$) for the specific a's and c's in part (e) by checking equation (1) for exactly one more function.

84. Suppose a student's social security number is $abc - de - fghi$. I can take the digits $a, b, c, d, e, f, g, h, i$ and $1, 2, 3, 4, 5, 6, 7$ and form the matrix

$$A = \begin{bmatrix} a & b & c & d \\ e & f & g & h \\ i & 1 & 2 & 3 \\ 4 & 5 & 6 & 7 \end{bmatrix}.$$

Do this for your own social security number, to form your own special matrix.

As the instructor, with MATLAB, I save all the students' special matrices, which I can pull up for extra-credit quizzes during the semester. I might ask the students to find the row reduced echelon form of their matrix, LU factorization, QR factorization, inverse if it exists (there is a high probability it does exist), determinant, the largest eigenvalue and the corresponding eigenvector, and so on.

I give such exercises as extra credit: $+5$ points if correct; -5 points if there is a single error; no partial credit. They have the option of doing all, some, or none. Few students have MATLAB, they have to work through the details carefully and I ask for exact answers whenever I can. These extra credit problems both generate interest among students (for the extra credit) and cause them to think about being careful in their work.

Jim Weaver
University of North Florida

85. An economy of two industries has the following I/O matrix A and demand matrix D. Apply the Leontief model to this situation. Why does it work? Explain the situation from both a mathematical and a practical viewpoint. Extend your results to the general model.

$$A = \begin{bmatrix} .08 & .06 \\ .02 & .04 \end{bmatrix}, \qquad D = \begin{bmatrix} 20 \\ 32 \end{bmatrix}$$

Gareth Williams
Stetson University

86. Markov chain models for predicting future population distributions are standard applications in linear algebra courses. The following questions are interesting and are not usually given.

   Annual population movement between U.S. cities and suburbs in 1985 is described by the following stochastic matrix P. The population distribution in 1985 is given by the matrix X, in units of one million.

$$
\begin{array}{cc}
 & \text{(from)} \\
 & \text{City\ \ Suburb\quad (to)}
\end{array}
$$
$$
P = \begin{bmatrix} 0.96 & 0.01 \\ 0.04 & 0.99 \end{bmatrix} \begin{array}{l} \text{City} \\ \text{Suburb} \end{array} \qquad X = \begin{bmatrix} 60 \\ 125 \end{bmatrix} \begin{array}{l} \text{City} \\ \text{Suburb} \end{array}
$$

   a) Determine the population distributions for 1980 to 1985—prior to 1985. Is the chain going from 1985 to the present a Markov chain? What are the characteristics of the matrix that takes one from distribution to distribution? Give a physical reason for these characteristics.

   b) Interchange the columns of P to get a stochastic matrix Q. Compare the Markov chains for stochastic matrices P and Q. Which model is most realistic?

   Gareth Williams
   Stetson University

Hermitian Matrices

87. Let $m \geq n$ and let A be a complex $m \times n$ matrix of rank n. Let A^* denote the conjugate transpose of A. Then the Gram matrix $G = A^*A$ is a positive definite Hermitian $n \times n$ matrix. Show that the Hermitian matrix $B = I_m - AG^{-1}A^*$ is nonnegative if I_m denotes the $m \times m$ unit matrix.

   Hans Schwerdtfeger
   McGill University
   soln [1963, 902]

88. It is well known that for any square matrix A with complex elements, there is a unique decomposition $A = B + C$, where $B = (A + A^*)/2$ is Hermitian (and has all roots along the real axis) and $C = (A - A^*)/2$ is skew-Hermitian (and has all roots along the imaginary axis). Given any two distinct lines through the origin in the complex plane, prove an analogous result for a unique decomposition of A into two "Hermitian-like" matrices, each with roots along one of the two lines.

   David Carlson
   Oregon State University
   soln [1965, 325]

89. Let

$$X = \begin{pmatrix} X_{11} & \cdots & X_{1n} \\ \vdots & & \vdots \\ X_{n1} & \cdots & X_{nn} \end{pmatrix}$$

denote the Hermitian, positive definite matrix, where each block of the partitioning is $m \times m$, and define

$$Y = \begin{pmatrix} Y_{11} & \cdots & Y_{1n} \\ \vdots & & \vdots \\ Y_{n1} & & Y_{nn} \end{pmatrix} = \begin{pmatrix} X_{11} & \cdots & X_{1n} \\ \vdots & & \vdots \\ X_{n1} & & X_{nn} \end{pmatrix} = X^{-1}$$

with corresponding partitions. Prove that $\left(\sum_{r,t=1}^{n} Y_{rt} \right)^{-1}$ is Hermitian, positive definite.

H.J. Thiebaux
Boulder, Colorado
soln [1976, 388]

90. Let A and B be nonnegative definite Hermitian matrices such that $A - B$ is also nonnegative definite. Show that $tr(A^2) \geq tr(B^2)$.

Götz Trenkler
University of Dortmund, Germany
prob 10234 [1992, 571]

91. Let A be an n-square Hermitian matrix whose characteristic roots are the diagonal elements $A_{ii}, i = 1, \cdots, n$. Prove that A is diagonal.

M.D. Marcus and N. Moyls
University of British Columbia
soln [1956, 124]

92. Show that a matrix which is similar to a real diagonal matrix is the product of two Hermitian matrices one of which is positive definite. (The converse is known.)

Olga Taussky
California Institute of Technology
soln [1960, 192–193]

93. Let A, B be two positive definite Hermitian matrices which can be transformed simultaneously by unitary transformation to diagonal forms of similarly ordered numbers. Let x be any vector of complex numbers. Show that $(Ax, x)(Bx, x) \leq (ABx, x)(x, x)$, and discuss the case of equality. Two sets of n positive numbers, $\{a_i\}, \{b_i\}$ are called similarly ordered if $(a_i - a_k)(b_i - b_k) \geq 0$ for all $i, k = 1, \cdots, n$.

Olga Taussky
California Institute of Technology
soln [1961, 185-186]

Magic Squares

94. Prove that the matrix product of any two third-order magic squares is a doubly symmetric matrix. A magic square is defined as a square array of n^2 elements with the property that the sum of the elements of any line (both diagonals included) is a constant; double symmetry is defined as symmetry in both diagonals.

G.P. Sturm
Oklahoma State University
soln [1962, 65]

95. Let A be a row semi-magic square and let A_j denote the matrix obtained by replacing the jth column of A by a column of 1's. Show that $\det(A_j)$ is independent of j.

Milton Legg
University of Minnesota
soln [1967, 200]

96. Let A be a 3 by 3 magic matrix with real elements; i.e., there is a nonzero real number s such that each row of A sums to s, each column of A sums to s, the main diagonal of A sums to s, and the counter-diagonal of A sums to s.

(a) Show that if A is also nonsingular, then A^{-1} is magic.

(b) Show that A has the form

$$\begin{pmatrix} s/3+u & s/3-u+v & s/3-v \\ s/3-u-v & s/3 & s/3+u+v \\ s/3+v & s/3+u-v & s/3-u \end{pmatrix},$$

where u and v are arbitrary, and nonsingular if and only if $v^2 \neq u^2$.

William P. Wardlaw
U.S. Naval Academy, Annapolis, MD
soln [1992, 966]

97. A magic matrix is one whose elements are the numbers of a magic square, i.e., every row, column, and diagonal has the same sum. (a) Show that a 3 by 3 magic matrix inverts into a magic matrix. (b) Can this result be extended to magic matrices of higher order?

Charles Fox
McGill University
soln [1957, 599]

98. Show that a semi-magic matrix A (the sums of the rows and columns are all equal) can be decomposed into a sum $B + C$ such that for integral K,

$$(B+C)^K = B^K + C^K.$$

F.D. Parker
University of Alaska
soln [1960, 703]

Special Matrices

99. Find the most general square root of the 3×3 identity matrix if the elements are to be (a) integers, (b) any real numbers, (c) complex numbers.

   George Grossman
   Board of Education, New York
   soln [1969, 303]

100. Let A be an $n \times n$ matrix with entries zero and one, such that each row and each column contains precisely k ones. A *generalized diagonal* of A is a set of n elements of A such that no two elements appear in the same row or the same column. Show that A has at least k pairwise disjoint generalized diagonals, each of which consists entirely of ones.

   E.T.H. Wang
   University of Waterloo, Canada
   soln [1973, 945–946]

101. Find all matrices A such that both A and A^{-1} have all elements real and nonnegative.

   H. Kestelman
   University College, London, England
   soln [1973, 1059–1060]

102. Characterize all $n \times n$ complex matrices A for which the relation $\text{per}(AX) = \text{per}(A)\text{per}(X)$ holds for every complex $n \times n$ matrix X. (Here $\text{per}(B)$ denotes the permanent of B.) Does the same characterization hold when the complex field is replaced by an arbitrary field?

   Ko-Wei Lih
   Academia Sinica, Taiwan
   soln [1982, 605–606]

103. A square matrix of order $n, n \geq 2$, is said to be "good" if it is symmetric, invertible, and all its entries are positive. What is the largest possible number of zero entries in the inverse of a "good" matrix?

   Miroslav D. Asič
   London School of Economics
   soln [1986, 401–402]

104. Let S be an m by m matrix over $\mathbb{C}$. It is well known that $S^2 = S$ and $\text{trace}(S) = 0$ imply that S is the zero matrix. For which positive integers $n > 2$ and $m > 1$ does the pair of conditions $S^n = S$ and $\text{trace}(S) = 0$ imply that S is the zero matrix?

   David K. Cohoon
   Temple University
   soln [1989, 448–449]

105. (a) Prove that a (square) matrix over a field F is singular if and only if it is a product of nilpotent matrices.

(b) If $F = \mathbb{C}$, prove that the number of nilpotent factors can be bounded independently of the size of the matrix.

Daniel Goffinet
St. Étienne, France
soln [1993, 807–809]

106. Suppose that A is an n by n matrix with rational entries whose multiplicative order is 15; i.e. $A^{15} = I$, an identity matrix, but $A^k \neq I$ for $0 < k < 15$. For which n can one conclude from this that

$$I + A + A^2 + \cdots + A^{14} = 0?$$

William P. Wardlaw
United States Naval Academy, Annapolis, MD
prob 10318 [1993, 590]

107. Let $S_1, S_2, \ldots, S_k$ be a list of all non-empty subsets of $\{1, 2, \ldots, n\}$. Thus $k = 2^n - 1$. Let $a_{ij} = 0$ if $S_i \cap S_j = \phi$ and $a_{ij} = 1$ otherwise. Show that the matrix $A = (a_{ij})$ is non-singular.

Anthony J. Quinzi
Temple University
soln [1979, 308]

Stochastic Matrices

108. Show that every normal stochastic matrix is necessarily doubly stochastic.

Richard Sinkhorn
University of Houston
soln [1971, 547]

109. A square matrix is *doubly-stochastic* if its entries are non-negative and if every row sum and every column sum is one. Show that every doubly-stochastic matrix (other than the one with all entries equal) contains a 2×2 submatrix

$$\begin{pmatrix} a & b \\ c & d \end{pmatrix}$$

such that either $\min(a, d) > \max(b, c)$ or $\max(a, d) < \min(b, c)$.

Franz Hering
University of Washington
soln [1973, 200]

110. Let A be a row stochastic matrix such that $\|A\| = 1, \| \ \|$ being the operator norm induced by the Euclidean vector norm. Show that A is doubly stochastic.

Emeric Deutsch
Polytechnic Institute of New York
soln [1983, 409–410]

Trace

111. For square matrices of order n, prove that any matrix M: (1) has zero trace if $MQ = -QM$, where Q is some nonsingular matrix, (2) can be written as a matrix of zero trace plus a multiple of the identity.

    R.G. Winter
    Pennsylvania State University
    soln [1962, 436–437]

112. If A is a 3×3 involutoric matrix $(A^2 = 1)$ with no zero elements, prove that the trace of A is $+1$ or -1.

    R.E. Mikhel
    Ball State Teachers College
    soln [1964, 436]

113. Let A be an $n \times n$ matrix, $A^2 = I$, the identity, and $A \neq \pm I$. Show

$$(1)\ Tr(A) \equiv n(\mathrm{mod}\ 2), \qquad (2)\ |Tr(A)| \leq n - 2,$$

    where $Tr(A)$ is the trace of A.

    H.A. Smith
    Institute for Defense Analysis
    soln [1967, 1277–78]

114. Let $a_k = \sum_{i=1}^{n}(x_i)^k, k = 1, 2, \cdots$. Show that $a_k(k = n + 1, n + 2, \cdots)$ can be expressed as a polynomial in $a_1, a_2, \cdots, a_n$. Stated differently, show that the traces of the higher powers of an $n \times n$ matrix can be expressed as polynomials in the traces of the first n powers.

    P.J. Schweitzer
    Institute for Defense Analyses
    soln [1968, 800]

Other Topics

115. Let V and W be two vector spaces over the same field. Suppose f and g are two linear transformations $V \to W$ such that for every $x \in V, g(x)$ is a scalar multiple (depending on x) of $f(x)$. Prove that g is a scalar multiple of f.

    Edward T.H. Wang and Roy Westwick
    Wilfri Laurier University and University of British Columbia
    soln [1981, 348–349]

Problems 116-121, 123 submitted by Steve Maurer, Swarthmore

116. If the point of the definition of Span(S) is to augment S just enough to get a vector space, why don't we define Span (S) to be

$$\{c\mathbf{v}|c \in R, \mathbf{v} \in S\} \cup \{\mathbf{u} + \mathbf{v}|\mathbf{u}, \mathbf{v} \in S\}?$$

If that's no good, why not

$$\{c\mathbf{u} + d\mathbf{v}|c, d \in R, \mathbf{u}, \mathbf{v} \in S\}?$$

117. Devise and justify an algorithm for determining whether the span of one set of vectors is contained in the span of another.

118. Call a set $\{\mathbf{v}_1, \ldots, \mathbf{v}_k\} \subset R^n$ *strongly dependent* if there exists some set of constants, *none* 0, so that $\sum_{i=0}^{k} c_i\mathbf{v}_i = \mathbf{0}$.

   a) How does strong dependence differ from dependence?

   b) Prove that if $S = \{\mathbf{v}_1, \ldots, \mathbf{v}_k\}$ is strongly dependent, then every $\mathbf{v}_i \in S$ depends on $S - \{\mathbf{v}_i\}$.

   c) Is the converse of part b true? (Give a proof or counterexample.)

   I made up this definition of strong dependence. It doesn't have much use except to help you understand (ordinary) dependence better.

119. Suppose $\mathbf{v}_1, \mathbf{v}_2, \ldots, \mathbf{v}_k$ are in vector space V. State what is meant for $\{\mathbf{v}_1, \ldots, \mathbf{v}_k\}$ to be a basis of V

   a) using the words "span" and "independent";

   b) using instead the phrase "linear combination."

120. Name specific vectors $\mathbf{u}, \mathbf{v}$, and $\mathbf{w}$ in R^2 that satisfy all the following conditions:

   a) they are independent;

   b) $\mathbf{u}$ and $\mathbf{v}$ are independent;

   c) $\mathbf{v}$ does not depend on $\mathbf{u}$ and $\mathbf{w}$.

121. All of the following statements are incorrect because mathematical words are abused (not because the intended meanings are wrong). Correct each statement by changing or adding a few words or by restating the whole sentence.

   a) A matrix A is invertible if it has unique solutions.

   b) Every basis of a vector space V has the same dimension.

   c) Vectors $\mathbf{u}, \mathbf{v}$ are orthogonal complements if $u^T v = 0$.

   d) To get a matrix A with $R(A) = V$, let the rows of A be the span of V.

   e) A nontrivial linear combination of vectors is independent if it does not equal 0.

   f) The number of vectors in $R(A)$ is the number of free variables of A.

   g) An independent set $\{v_1, v_2, \ldots, v_n\}$ is the dimension of V if it is the span.

122. Give an example of a subset of $\mathbf{R}^3$ that is

    a) closed under addition, but not closed under scalar multiplication;

    b) closed under scalar multiplication, but not closed under addition.

    Such examples illustrate the independence of these two conditons.

    Gareth Williams
    Stetson University

123. Consider a set of k vectors in R^n and ask the following three questions: Are they linearly independent? Do they span R^n? Do they form a basis? The answers can be listed in a table as shown.

	$k < n$	$k = n$	$k > n$
Indep	Probably yes	Yes or no	Never
Basis	Never	Yes or no	Never
Span	Never	Yes or no	Probably yes

    The "never" answers are clear. The "probably yes" answers are illustrated geometrically in low dimensions. (Students really love this "probably" in an otherwise very dry and abstract course.) The point about the "yes or no" is that the answer is the same to all three questions.

AUTHOR INDEX

FURTHER ITEMS OF INTEREST

Our search of Mathematical Association of America publications turned up many other items that we have not been able to include. For the interested reader, references to a selection of these items follows. All are from the *American Mathematical Monthly* (*AMM*) or the *College Mathematics Journal* (*CMJ*).

Row Reduction of a Matrix and CaB, Stuart L. Lee, and Gilbert Strang, *AMM* **107** (2000), 668–688.

Determinants of Commuting Block Matrices, Istvan Kovacs, Daniel S. Silver, and Susan G. Williams, *AMM* **107** (2000), 950–952.

Eigenvalues of Matrices of Low Ranks, Stewart Venit and Richard Katz, *CMJ* **31** (2000), 208–210.

Collapsed Matrices with (Almost) the Same Eigenstuff, Don Hooley, *CMJ* **31** (2000), 297–299.

Using Consistency Conditions to Solve Linear Systems, Geza Schay, *CMJ* **30** (1999), 226–229.

The Rook on the Half Chessboard, or How Not to Diagonalize a Matrix, Kiran S. Kedlaya and Lenhard L. Ng, *AMM* **105** (1998), 819–824.

Separation of Subspaces by Volume, David W. Robinson, *AMM* **105** (1998), 22–27.

Finding a Determinant and Inverse by Bordering, Y.-Z. Chen and R.F. Melka, *CMJ* **29** (1998), 38–39.

A Diagonal Perspective on Matrices, Eugene Boman, *CMJ* **29** (1998), 37–38.

When Is Rank Additive?, David Callan, *CMJ* **29** (1998), 145–147.

Clock Hands Pictures for 2-by-2 Real Matrices, C.R. Johnson and B.K. Kroschel, *CMJ* **29** (1998), 148–150.

A Fresh Approach to the Singular Value Decomposition, C. Mulcahy and J. Rossi, *CMJ* **29** (1998), 199–207.

Generating Exotic Looking Vector Spaces, Michael Carchidi, *CMJ* **29** (1998), 304–308.

When Is a Linear Operation Diagonalizable?, Marco Abate, *AMM* **104** (1997), 824–830.

Some Inequalities for Principal Submatrices, John Chollet, *AMM* **104** (1997), 609–617.

Hereditary Classes of Operators and Matrices, Scott A. McCullough and Leiba Rodman, *AMM* **104** (1997), 415–430.

Applications of Linear Algebra in Calculus, Jack W. Rogers, Jr., *AMM* **104** (1997), 20–26.

The Generalized Spectral Decomposition of a Linear Operator, Garret Sobczyk, *CMJ* **28** (1997), 27–38.

Eigenpictures and Singular Values of a Matrix, Peter Zizler, *CMJ* **28** (1997), 59–62.

K-volume in R^n and the Generalized Pythagorean Theorem, Gerald J. Porter, *AMM* **103** (1996), 252–256.

A Singularly Valuable Decomposition: The SVD of a Matrix, Dan Kalman, *CMJ* **27** (1996), 2–23.

Complex Eigenvalues and Rotations: Are Your Students Going in Circles?, James Deummel, *CMJ* **27** (1996), 378–381.

The Ranks of Tournament Matrices, T.S. Michael, *AMM* **102** (1995), 637–639.

Matrix Expansion by Orthogonal Kronecker Products, Jeffrey C. Allen, *AMM* **102** (1995), 538–540.

A Simple Estimate of the Condition Number of a Linear System, H. Guggenheimer, A. Edelman and C.R. Johnson, *CMJ* **26** (1995), 2–5.

Eigenpictures: Picturing the Eigenvector Problem, S. Schonefeld, *CMJ* **26** (1995), 316–319.

EDITORS' BIOGRAPHICAL INFORMATION

Dave Carlson has been a Professor of Mathematics at San Diego State University since 1982. He received a B.A. in 1957 at San Diego State College, a M.S. in 1959 and his Ph.D.(in matrix theory under Hans Schneider) in 1963 at the University of Wisconsin. Dr. Carlson's professional interests include linear algebra and matrix theory. He especially enjoys teaching linear algebra. He is an advisory editor of *Linear Algebra and Its Applications*, an organizer of the Linear Algebra Curriculum Study Group, and a member of the International Program Committee for the ICMI Study on the Future of the Teaching and Learning of Algebra.

Charles Johnson received his B.A. (Northwestern University) and his Ph.D (California Institute of Technology, 1972) in both mathematics and economics. He was an NAS-NRC postdoc at the National Bureau of Standards (now NIST) before moving to the University of Maryland in 1974, where he was tenured in 1976 in the Institute for Fluid Dynamics and Applied Mathematics (later IPST) and the Economics Department. After a brief tenure at Clemson University, he went to the Mathematics Department at the College of William and Mary in 1987 as Class of 1961 Professor (an endowed chair) and Virginia Eminent Scholar. He remains at William and Mary where he has built the world's most recognized group in matrix analysis and received the 2001 state-wide Virginia Outstanding Faculty Award. Johnson works primarily in matrix analysis and allied areas, such as combinatorics. He has published about 250 papers and several books, including the two volumes, *Matrix Analysis* and *Topics in Matrix Analysis* (with R. Horn). He has been a principle invited speaker at more than 150 meetings (including AMS, MAA and SIAM meetings) in a wide variety of areas, and given nine lecture series, most recently the 2001 Rocky Mountain Mathematics Consortium two-week, summer lecture series. He has served as an editor of all the major journals in linear algebra, as a managing editor of a *SIAM* journal, and as an editor of journals in allied areas. He has frequently taught linear algebra and has an active interest in how students assimilate the concepts of elementary linear algebra.

David C. Lay received his Ph.D at UCLA in 1966 and since then has been at the University of Maryland, with sabbatical leaves at several universities abroad. He has over 30 research articles published in functional analysis and linear algebra. He has been an invited speaker at both research conferences and MAA meetings. In 1999, he was a plenary speaker at the winter meeting of the Canadian Mathematical Society. He is a member of several professional societies, including the International Linear Algebra Society. He has been on the national board of the Association of Christians in the Mathematical Sciences, and he has served on CRAFTY, a subcommittee of the MAA. Lay is the author or co-author of nine books, including *Introduction to Functional Analysis* (with A.E. Taylor), *Calculus and Its Applications* (with L. Goldstein and D. Schneider, currently in its ninth edition), and *Linear Algebra and Its Applications* (currently in its second edition). Lay has received four university awards for teaching excellence, including the title of Distinguished Scholar-Teacher of the University of Maryland. University students have elected him to honorary membership in Alpha Lambda Delta National Scholastic Honor Society and Golden Key National Honor Society. In 1989, Aurora University conferred on him the Outstanding Alumnus award. In 1994, he received an MAA award for Distinguished College or University Teaching of Mathematics.

A. Duane Porter received his B.S. and M.S. degrees from Michigan State University (1960 and 1961). He received his Ph.D from the University of Colorado in 1964 with an emphasis in matrix theory and has been at the University of Wyoming since that time. Dr. Porter has published over 50 research articles in matrix theory and finite fields. In 1992 he was awarded the "Certificate of Meritorious Service" by the Rocky Mountain Section of the MAA. They also presented him with the "Award for Distinguished College or University Teaching of Mathematics" in 1994. The University of Wyoming selected him as the recipient of the Duke Humphrey Award in 1998, the university's highest honor. His professional interests include algebra and linear algebra, the teaching of linear algebra and working with secondary teachers to improve public school mathematics. He has been active in many professional organizations, having served on the MAA Board of Governors, as a Section Chair for two terms, and as a founding member of the Linear Algebra Curriculum Study Group. He has been a member of the Board of Directors of the Rocky Mountain Mathematics Consortium for 30 years and serves as a linear algebra editor for the Rocky Mountain Journal of Mathematics.